MALDI-TOF Mass Spectrometry in Microbiology

Edited by

Markus Kostrzewa

Bruker Daltonik GmbH
Bremen
Germany

and

Sören Schubert

Max von Pettenkofer-Institut
Munich
Germany

Caister Academic Press

Caister Academic Press
Norfolk, UK

www.caister.com

British Library Cataloguing-in-Publication Data
A catalogue record for this book is available from the British Library

ISBN: 978-1-910190-41-8 (paperback)
ISBN: 978-1-910190-42-5 (ebook)

Cover design adapted from images courtesy of Bruker Daltonics Inc.

Contents

Contributors

Merima Alispahic
Clinic for Poultry and Fish Medicine
University of Veterinary Medicine Vienna
Vienna
Austria

amerima@gmail.com

Silpak Biswas
Faculty of Medicine and Pharmacy
IHU Méditerranée Infection
Aix-Marseille University
Marseille
France

silpakbiwas@gmail.com

Chantal Bizet
Institute Pasteur
Collection de l'Institut Pasteur; and
Centre de Ressources Biologiques de l'Institut Pasteur
Microbiology Department
Paris
France

chantal.bizet@pasteur.fr

Teun Boekhout
CBS Fungal Biodiversity Centre (CBS-KNAW)
Utrecht
The Netherlands; and
Institute of Microbiology
Chinese Academy of Sciences
Beijing; and
Department of Dermatology
Shanghai Key Laboratory of Molecular Medical Mycology
Institute of Dermatology and Medical Mycology
Changzheng Hospital
Second Military Medical University
Shanghai
China

t.boekhout@cbs.knaw.nl

Dominique Clermont
Institute Pasteur
Collection de l'Institut Pasteur
Microbiology Department
Paris
France

dominique.clermont@pasteur.fr

Christopher D. Doern
Virginia Commonwealth University Medical Center
Medical College of Virginia Campus
Richmond, VA
USA

christopher.doern@vcuhealth.org

Monika Dolejská
Biomedical Center
Faculty of Medicine and University Hospital in Plzen
Charles University in Prague
Plzen; and
Department of Wildlife Diseases
Faculty of Veterinary and Ecology
Central European Institute of Technology
University of Veterinary and Pharmaceutical Sciences Brno
Brno
Czech Republic

dolejskam@vfu.cz

Mark D. Gonzalez
Children's Healthcare of Atlanta
Emory University School of Medicine
Atlanta, GA
USA

mark.gonzalez@choa.org

Frédérique Gouriet
Faculty of Medicine and Pharmacy
IHU Méditerranée Infection
Aix-Marseille University
Marseille
France

frederique.gouriet@univ-amu.fr

Claudia Hess
Clinic for Poultry and Fish Medicine
University of Veterinary Medicine Vienna
Vienna
Austria

claudia.hess@vetmeduni.ac.at

Michael Hess
Clinic for Poultry and Fish Medicine
University of Veterinary Medicine Vienna
Vienna
Austria

michael.hess@vetmeduni.ac.at

Christiane Honisch
Illumina, Inc.
San Diego, CA
USA

chonisch@illumina.com

Jaroslav Hrabák
Biomedical Center
Faculty of Medicine in Plzen; and
Department of Microbiology
Faculty of Medicine and University Hospital in Plzen
Charles University in Prague
Plzen
Czech Republic

jaroslav.hrabak@lfp.cuni.cz

Robert C. Jerris
Children's Healthcare of Atlanta
Emory University School of Medicine
Atlanta, GA
USA

robert.jerris@choa.org

Anna Kolecka
CBS Fungal Biodiversity Centre (CBS-KNAW)
Utrecht
The Netherlands

a.kolecka@cbs.knaw.nl

Markus Kostrzewa
Bruker Daltonik GmbH
Bremen
Germany

markus.kostrzewa@bruker.com

Elizabeth Nagy
Institue of Clinical Microbiology
University of Szeged
Szeged
Hungary

nagy.erzsebet@med.u-szeged.hu

Costas C. Papagiannitsis
Biomedical Center
Faculty of Medicine in Plzen; and
Department of Microbiology
Faculty of Medicine and University Hospital in Plzen
Charles University in Prague
Plzen
Czech Republic

papagiac@ifp.cuni.cz

Brunella Posteraro
Institute of Public Health (Section of Hygiene)
Catholic University of Rome
Rome
Italy

brunella.posteraro@unicatt.it

Jean-Marc Rolain
Faculty of Medicine and Pharmacy
IHU Méditerranée Infection
Aix-Marseille University
Marseille
France

jean-marc.rolain@univ-amu.fr

Maurizio Sanguinetti
Institute of Microbiology
Catholic University of Rome
Rome
Italy

maurizio.sanguinetti@unicatt.it

Sören Schubert
Max von Pettenkofer-Institut
Munich
Germany

schubert@med.uni-muenchen.de

Haroun N. Shah
Department of Natural Sciences
Middlesex University
Middlesex

harounshah@gmail.com

Preface

Microbiology is thought to be a very conservative area in science and medicine with clinical microbiology being its most conservative part. Still Gram-staining and microscopy are central tools in the microbiology laboratory like a century ago. But sometimes revolutions change the game. Such a revolution has taken place when MALDI-TOF mass spectrometry has entered microbiological practice. Less than a decade has passed since the first routine users started to express their excitement about the benefits they gained with the new technology, about increased accuracy for the broad spectrum of nearly all microorganisms appearing in daily practice, tremendously shortened time to get an identification result, and significantly lowered overall costs. While the technology started its success story in Europe, it quickly also entered the Americas as well as Asian countries and now is getting adapted there.

Clinical diagnostic was the first area where MALDI-TOF MS was introduced for broad spectrum microorganism identification in routine practice. Both direct impact on healthcare and economic advantages were powerful drivers helping to overcome conservative reservation. But other fields quickly followed, from scientific research in different areas to industrial quality control. The spectrum of organisms which were investigated using the technology permanently expanded. Today, MALDI-TOF MS is widely accepted as the new laboratory standard for microorganism identification, and further application fields are on the horizon. As nowadays the identification of organisms with lowest preparation efforts is daily business in many laboratories, many microbiologists are looking for more applications which might benefit from the capabilities of such a system. This is in fact supported by the technological development on instrument and software side. The instruments as well as analysis systems are becoming faster, thereby not only results are available even earlier but also time is freed for additional assays.

We both participated in the revolution in microbiology caused by MALDI-TOF MS, as technology developer and early adaptor in a routine laboratory. We have experienced the scientific wave which it has caused during the recent years but also the enormous impact on practice in clinical microbiology. Still we are excited about the impact of this technology but we also think there is no reason to stand still. In this book experts in the field describe the state-of-the-art and give an outlook on how MALDI-TOF MS may continue the revolution. This shall be a help for the unexperienced reader but also an inspiration for scientists and practitioners to participate.

Finally, we would like to commemorate the co-inventor of the MALDI technique, Franz Hillenkamp. In the late 1980s he 'saw the first ions fly' with his invention. We experienced his great excitement when he got aware about the introduction of 'his' technology into a clinical field, some time after his retirement. He passed away before he could write the planned chapter for this book. Scientists with his brightness and enthusiasm are rare and urgently needed for revolutions as caused by MALDI-TOF MS in microbiology.

In memoriam Franz Hillenkamp
Markus Kostrzewa
Sören Schubert

Introduction

Haroun N. Shah

A personal vision of the MALDI-TOF-MS journey from obscurity to frontline diagnostics

Applied science is driven by advances in technology, particularly in microbiological research, which has been the recipient of huge strides in genome and proteome-based technologies. However, the path from technological development to specific application/s is often skewed and unpredictable. New technologies inspire scientific investigations but, if results are inconclusive, interest soon declines. By contrast, a glint of hope can act as a catalyst and result in an upsurge in scientific publications, frequently based upon limited datasets generated in a few laboratories. Consequently, success and eventual implementation is dependent on a number of factors; prominent among them being the willingness of scientists to embrace the change that will most likely affect the workflow of their laboratory. Over the years microbiological applications of mass spectrometry have followed both paradigms.

The uptake of MALDI-TOF mass spectrometry (MALDI-TOF MS) by clinical laboratories has been phenomenal, yet its arrival into microbiology could have taken the same course as earlier forms of mass spectrometry which experienced brief periods of success, followed by the inevitable obscurity as in the case of microbial applications of Pyrolysis Mass Spectrometry. However, in the case of MALDI-TOF MS, numerous factors have contributed to its success and laid the groundwork for its universal acceptance as a fundamental platform for diagnostic microbiology.

During my career at the microbiological laboratories of the Public Health Laboratory Service [PHLS, subsequently Health Protection Agency in 2002, and from 2013 as Public Health England (PHE)], I was involved in research to explore and develop mass spectrometry-based applications in clinical microbiology. This engagement was at the time when there was a climate of fierce resistance both from within PHLS and external peer reviewers who were aware of the earlier failures of mass spectrometry to secure a foothold in clinical microbiology. The concept of exploiting cellular proteins for microbial identification using charge and mass values as markers was a bizarre concept to the majority of clinical microbiologists. A decade of design, re-analysis and validation ushered the gradual acceptance of MALDI-TOF into diagnostic laboratories but the mainstream, reference and public health laboratories remained sceptical for a long period.

This 15-year journey may therefore serve as a reference for technological transformation which often requires the dogged persistence of a few scientists who feel confident and zealous to commit to a new innovation and are prepared to challenge the core foundation of their institute to establish it. This becomes even more acute when the technology requires considerable development as in the case of MALDI-TOF MS at its inception. In 1996, four publications Claydon *et al.* (1996), Cain *et al.* (1996), Holland *et al.* (1996) and Krishnamurthy *et al.* (1996) highlighted the potential of MALDI-TOF MS to discriminate between fairly disparate species. My interest began as I exchanged views and aspirations with M.A. Claydon in late 1997. I had just left the University of London to join the then PHLS to create a new Molecular Identification Services Unit (MISU) whose remit was the characterization of atypical, rarely isolated and emerging pathogens. MISU was established using mainly 16S rRNA and long-chain cellular fatty acid analysis and I was in search of new approaches to characterize very unusual clinical isolates that were received from a variety of clinical and environmental sources.

My prior research experience on electron impact mass spectrometry was a basis for my interest in new forms of mass spectrometry and its potential for differentiating bacteria. In fact in the 1980s, using mass spectrometry to analyse non-polar lipids from some of the most difficult to identify anaerobic bacteria, was a cornerstone for chemotaxonomy and led to the taxonomic restructuring of one of the largest bacterial families, the *Bacteroidaceae* (Shah and Collins, 1980; Shah and Gharbia, 2011). Although the MS technology then had an upper mass limit to analyse compounds < 1.5 kDa, the impact of comparing cellular components such as quinones and porphyrins was remarkable and informed the taxonomy and discovery of new taxa and species across the microbial kingdom. When MALDI-TOF MS emerged in the 1990s, in my view, it offered considerably more potential since it had the capacity to analyse significantly larger molecules (> 30 kDa) and I had unfathomable confidence in the enormity of the shift in the applications of mass spectrometry.

As applications of MALDI-TOF MS in microbiology were presented at meetings and published, microbiologists remained highly distrustful largely because earlier attempts to introduce various methods such as Pyrolysis MS, Electron Impact MS and Fast Atomic Bombardment MS endured minimal success. It was against this background that we set out to methodically develop the diagnostic applications of MALDI-TOF MS between 1998 and 2012 through extensive work and collaboration with industry and academic partners, during which a continuous stream of 15 grant-funded PhD students were trained and contributed to eventually establishing a validated working system.

Together with Kratos Analytical and Manchester Metropolitan University, we successfully organized, in 1998, the first of a series of annual international conferences entitled 'Intact Cell MALDI'. My presentation entitled, 'A review of the current methods of bacterial identification and MALDI-TOF MS in the characterization of the obligate anaerobes *Fusobacterium* and *Porphyromonas*' (H.N. Shah) heralded this early work as 'proof of principle' since prior to MALDI-TOF MS, the unambiguous delineation of these poorly defined, non-fermentative taxa was fraught with difficulties. By contrast, using MALDI-TOF MS, speciation was achieved within minutes using just a single colony (Shah, 2005; Shah *et al.*, 2002). However, while there were intermittent reports in the scientific community on the analysis of specific taxa there was no generally accepted method nor coherent plan to establish a universal approach to using MALDI-TOF MS as a diagnostic tool.

To my knowledge, we installed the first MALDI-TOF mass spectrometer in a clinical identification laboratory in the UK in 1998. This was in MISU and done in collaboration with Kratos Analytical, Manchester, where we undertook considerable fundamental work to establish standard protocols for sample processing, while assessing the safety requirements and risk of analysing whole bacterial cells. In particular spore-formers such as *Geobacillus stearothermophilus,* containing labelled antibiotic-resistant markers, were placed in the sample chamber of the instrument and numerous suction tubes were used to collect samples from potential vents to test for microbial viability and ensure that no cross-infection or contamination occurred. This work was published as an 'Application Note for Microbiology'. This 1-year programme of work highlighted areas for instrument improvement and were reported to the manufacturer. Among these were:

1 The horizontal positioning of the 'Time of Flight Tube' in the Kratos Kompact Alpha MS made the instrument too bulky for a microbiology laboratory especially at that time when clinical laboratories were shrinking in size as the transition to molecular-based methods, with its small, compact instruments, were being rapidly deployed.
2 The instrument needed to be redesigned with an upright flight tube to make sample loading more convenient and compatible with other laboratory instruments.
3 Replace the fragile 20-well target plate of the instrument by a more durable 96-well plate. Suggested inclusion of central wells for protein standards so that lock mass corrections could be made.
4 The instrument was very manual. We strongly advised more automation for sample analysis etc., of which automated laser energy setting was considered the most important. This was manually adjusted and took considerable time to optimize for a maximum mass density spectrum for each sample. A fixed laser energy setting was recommended.
5 The absolute need to develop a quality controlled database containing mass spectral profiles of type and reference strains representative of clinically important species.
6 The need for considerable software development for comparative analysis of strains and a prerequisite for a search engine to match the spectrum of an unknown to one in the database. To compare spectra from various samples initially, a modified Jaccard coefficient and UPGMA was used to analyse and interpret data which was also presented schematically as a dendrogram (see Shah *et al.*, 2000).

The next phase of development involved a long term collaboration with the mass spectrometry company, Micromass UK Ltd, Manchester and Manchester Metropolitan University (MMU) in 2001. Micromass designed and developed an instrument designated 'M@LDI' that incorporated all the features described above with a bespoke software 'MicrobeLynx'. Together with MMU, we secured funding for five years for the microbial database development. I was a co-member of the National Collection of Type Cultures (NCTC), and procured the entire bacterial collection to develop the foundation for a taxonomically verified database. The first year of the project was used to meticulously optimize protocols, interrogate the software and search engine. We undertook inter-laboratory reproducibility studies using three instruments in Manchester and London. Eighty strains were analysed monthly in replicates of twelve on each instrument and frequent meetings were held to assess the quality of the data and make improvements to the process. The stability of sample–matrix mixtures was evaluated over a time course of 12 months. Crucial to the entire project was the scrupulous assembly of the database. For this, the now fully operational M@LDI Linear Time of Flight Mass Spectrometer (Waters Corporation (Micromass) Ltd, Manchester, UK) was utilized. Operation of the mass spectrometer was performed using the MassLynxTM software. Automated calibration of the time of flight tube, followed by automated acquisition of the bacterial spectra was then performed using the real time data selection (RTDS) function in the MassLynxTM software. Spectral profiles were collected in the mass range 500–10,000Da, acquiring 10 shots per spectrum at a laser firing rate of 20 Hz. Fifteen spectra per sample well and 10 spectra per lock mass well were collected using the real time data selection option to optimize the collection of quality data. For database inclusion, the spectral reproducibility between the 12 replicates per sample was tested using a root mean square (RMS) calculation to identify and reject outliers at a value greater than 3.0. The RMS is the normalized deviation of the median of test spectra from the spectral average and therefore was used to compare each replicate spectrum in turn to the composite spectra of the remaining replicates. All verified spectra were combined to produce a composite spectral entry for each bacterium included in the database. Spectra from unknown isolates collected to test the database were processed and combined using the same parameters as for the database strains. The composite spectrum was then searched against the database using the MicrobeLynxTM software. Database searching was based upon an estimation of the probability that the mass spectral peaks in the test spectrum are comparable with the database spectrum. A list of the top matches was provided together with RMS values. A high probability and low RMS value indicated a good match.

Initially the instrument was set up to profile the surface molecules of cells; the rational being that differences between virulent versus avirulent strains, where the pathogenic potential is due to surface-associated molecules, could be mapped and used for detection of pathogenic variants. For some species such as *Peptostreptococcus micros* this was highly successful where resolution of the two pathotypes ('smooth' and 'rough' variants) were readily distinguishable through characteristic mass ions (Rajendram, 2003). However, to obtain such mass spectra, it was necessary to rigorously standardized parameters that may alter the morphology of cells grown on agar plates. The composition of the culture medium had a marked effect on colonial morphology and therefore yielded very different mass spectra for the same strain. Environmental factors such as pH, growth temperature, harvest time of cells, media constituents, selective agents, their source, atmospheric conditions, composition of various matrices etc. all affected the profile of the mass spectrum of a strain and needed to be separately evaluated to devise an optimum and reproducible protocol (see Shah *et al.*, 2000). Some media such as nutrient agar or selective media such as 'CLED' yielded very poor spectra. Based upon 100 species (50 of each Gram type), these parameters were rigorously tested. It was apparent that Columbia blood agar, with cultures grown between 24 and 48 hours provided mass spectra with the greatest density of mass ions and was used for creating the database of mass spectral profiles (Keys *et al.*, 2004). The database continued to be populated for a further year and then evaluated in a hospital environment. Of the 536 samples tested and compared against 16S rRNA and LCFA profiles, all species showed good congruence between MALDI-TOF MS and other methods (see Rajakaruna *et al.*, 2009). The exception was *Clostridium difficile* which yielded a confounding 90% failure of correct identification using this protocol. Several alternative matrices such as sinapinic acid, ferulic acid, 5-chloro-2-mercaptobenzothiazole etc. were unsuccessfully tried. Success was eventually achieved using 2,5-dihydroxybenzoic acid in acetonitrile–ethanol–water (1:1:1) with 0.3% TFA (DHB). Interestingly, electron microscopy (EM) revealed that while *C. difficile* cells were completely disintegrated when mixed with a cell suspension of DHB, cells smeared onto the MALDI-TOF target plate and viewed by scanning EM, showed only a loss of their outer polymeric layers, the rest of the cell remaining intact. The laser now readily penetrated the cell to profile the high abundance, conserved molecules such as the ribosomal proteins.

A linear MALDI-TOF MS does not have the capability to identify proteins, but by 2009 analysis of intracellular protein proved more consistent, because this new method targeted the abundant ribosomal proteins. Proof of this was obtained indirectly by using cellular fractions of *E. coli* (K12) prepared by ultracentrifugation and analysed by MALDI-TOF MS. Interestingly, the MALDI-TOF spectrum of the ribosomal fraction superimposed on the mass spectrum of whole cells. This was repeated using a range of species and similar results were obtained. This led to the re-examination of the matrices to enhance ribosomal protein ionization rather than surface molecules. Also, the concept of extracting bacterial cells with formic acid to improve the spectra was developed in collaboration with AnagnosTec GmbH, Potsdam/Golm. The result was a return to using alpha-cyano-4-hydroxycinnamic acid as a matrix solution for all bacterial taxa. Because the target molecules now moved from surface-associated to intracellular ribosomal RNA proteins, which are constantly present in the cell in high copy number, it was no longer necessary to culture cells under the very stringent regime done at the commencement of this project. This flexibility had a major bearing on the broader applications of the technology and was reported by the author at the European Society of Clinical Microbiology and Infectious Diseases, Vienna, 10–14 April 2010, a meeting considered a landmark in the arduous journey for acceptance of the technology for clinical laboratories.

From that fateful meeting of 1998 that focused solely on MALDI-TOF MS, by the following year, we began to use these annual international conferences as a springboard to showcase balanced presentations of advances in genomic and proteomic technologies. Over the last 18 years many eminent speakers, among them the late Emeritus Professor Franz Hillenkamp who coined the name, 'matrix-assisted laser-desorption/ionization time-of-flight mass spectrometry' to describe his discovery in the 1980s, delivered the keynote address, 'MALDI-TOF MS; Development of an Analytical Tool for Biological Sciences', to an exuberant audience in June 2008. While as early as 2002, Dirk van den Boom and Christiane Honisch (SEQUENOM) reported their success in establishing a MALDI-TOF method to sequence DNA in their presentation 'Applications of MALDI-TOF MS for Rapid DNA Typing'. Other applications of the technologies included using ProteinChip arrays followed by MALDI-TOF MS (designated SELDI) were presented by Ciphergen Biosystems Ltd. In 2007, Markus Kostrzewa presented Bruker's advances in microbial identification and typing and, in 2011 reported progress in database development and work on direct MALDI-TOF analysis of blood cultures. In 2012, Nikos Hontzeas reported bioMérieux's plans for automation in his presentation 'New horizons in the use of MALDI-TOF MS for the clinical microbiology laboratory'.

Consequently, through major industrial collaborations, we simultaneously investigated the potential of MALDI-TOF MS for microbial identification, DNA typing (see, for example, Bishop *et al.*, 2010) and biomarker discovery using the ProteinChip technology to pre-capture selected proteins prior to ionization (see review by Shah *et al.*, 2005). While MALDI-TOF MS is now firmly established and recognized globally, the author resolutely believes that the two latter approaches are immensely powerful technologies and their full potential for microbiology is grossly understated. Methods involving selective protein pre-capture, prior to MS, may help to refine the 'whole proteome' approach and while SELDI

ProteinChips are no longer manufactured, new methods involving magnetic bead capture and Bruker's AnchorChip may find wider application in the near future.

In hindsight, one can only speculate as to why MALDI-TOF MS succeeded while earlier forms of mass spectrometry failed. In our view, the major impetus for change came with the arrival of the first bench-top MALDI-TOF mass spectrometry (Kratos Analytical), which unlike its forerunners, was small, compact, practical, unintimidating and showed promise as early as 1996. However, to apply the rational to develop MALDI-TOF MS into a universally acceptable method that is capable of application across the microbial kingdom was a monumental task and required the sustained interest and commitment of its manufacturers. Changes in a company's focus determine success or failure of a new technology. On the near crest of success, Micromass (acquired by Waters Ltd) terminated the clinical microbiology database project; however, the momentum of MALDI-TOF MS was maintained by Bruker Daltonics Ltd and later by bioMérieux-Shimadzu. Much of the current success is owed to Bruker Daltonics Ltd who not only re-introduced the early instrument developed for the SEQUENOM mass cleave application, but worked steadfastly to develop a quality controlled database and the accompanying Biotyper software for simple usage. Kratos Analytical (acquired by Shimadzu) returned to the project by working informally with the small company, AnagnosTec GmbH (Am Mühlenberg, Potsdam/Golm who developed the SARAMIS software and database. We had worked closely with AnagnosTec between 2006 and 2010. AnagnosTec GmbH was acquired by bioMerieux, which announced its entry into the field at the ECCMID 2010 meeting in Vienna and built on the success of the SARAMIS software and associated comprehensive database. The lesser known company, SAI (Scientific Analysis Instruments, Manchester, UK), began work in this field almost simultaneously with Kratos Analytical and markets the MALDI-TOF MS Andromas with a strong base in France. For the foreseeable future, the role of MALDI-TOF MS in microbial identification appears assured. These companies have been marketing the technology on a global scale and is today changing the landscape of microbial diagnostics as witnessed in this book.

Although this sojourn for our laboratory began in 1998, the first commercial instrument (Bruker Microflex) was not installed for microbial identification in the Reference Microbiology Laboratories in England until November 2011; driven then by the need to have a simple, cost-effective, rapid and accurate method in place in the event of a major outbreak of infections at the London 2012 Olympics. Today, Public Health England's laboratories has perhaps the largest global network of instruments (> 15) in daily clinical use and results have gradually superseded phenotypic and molecular 16S rRNA sequencing methods which were the gold standard for the last decade. With European Union accreditation and the US Food and Drug Administration continuing to expand the list of approved species for MALDI-TOF MS clinical identification, it is likely that more mass spectrometry companies will join this revolution in diagnostics and current approaches and applications will expand even further.

References

Bishop, C., Arnold, C., and Gharbia, S.E. (2010). Transfer of a traditional serotyping system (Kauffmann-While) onto a MALDI-TOF MS Platform for the rapid typing of *Salmonella* isolates. In Mass Spectrometry for Microbial Proteomics, Shah, H.N., and Gharbia, S.E., eds. (Wiley, Chichester, UK), pp. 463–496.

Cain, T., Luberman, L.D., Weber, W.J. Jr., and Vertes, A. (1996). Differentiation of bacteria using protein profiles from matrix assisted laser desorption/ionization time-of-flight mass spectrometry. Rapid Comm. Mass Spectrometry *8*, 1026–1030.

Claydon, M.A., Davey, S.N., Edwards-Jones, V., and Gordon, D.B. (1996). The rapid identification of intact microorganisms using mass spectrometry. Nature Biotech. *14*, 1584–1586.

Holland, R.D., Wilkes, J.G., Rafii, F., Sutherland, J.B., Persons, C.C., Voorhees, K.J., and Lay, J.O. (1996). Rapid identification of intact whole bacteria based on spectral patterns using matrix-assisted laser desorption/ionization with time-of-flight mass spectrometry. Rapid. Comm. Mass Spectrometry *10*, 1227–1232.

Keys, C.J., Dare D.J., Sutton, H., Wells, G., Lunt, M., McKenna, T., McDowall, M., and Shah, H.N. (2004). Compilation of a MALDI-TOF mass spectral database for the rapid screening and characterisation of bacteria implicated in human infectious diseases. Infect. Genet. Evol. *4*, 221–242.

Krishnamurthy, T., Ross, P.L., and Rajamani, U. (1996). Detection of pathogenic and non-pathogenic bacteria by matrix-assisted laser desorption/ionization time-of-flight mass spectrometry. Rapid Comm. Mass Spectrometry *10*, 883–888.

Rajakaruna, L., Hallas, G., Molenaar, L., Dare, D., Sutton, H., Encheva, V., Culak, R., Innes, I., Ball, G., Sefton, A.M., *et al.* (2009). High-throughput identification of clinical isolates of *Staphylococcus aureus* using MALDI-TOF-MS of intact cells. Infect. Genet. Evol. *9*, 507–513.

Rajendram, D. (2003). Dissecting the Diversity of the Genus *Peptostreptococcus* using Genomics and Proteomic Analyses. PhD Thesis. (University of East London).

Shah, H.N. (2005). MALDI-TOF-Mass Spectrometry: Hypothesis to proof of concept for diagnostic microbiology. Clin. Lab. Intern. *29*, 35–38.

Shah, H.N., and Collins, M.D. (1980). Fatty acid and isoprenoid quinone composition in the classification of *Bacteroides melaninogenicus* and related taxa. J. Appl. Bacteriol. *48*, 75–84.

Shah, H.N., Encheva, V., Schmid, O., Nasir, P., Culak, R., Ines, I., Chattaway, M.A., Keys, C.J., Jacinto, R.C., Molenaar, L., *et al.* (2005). Surface Enhanced Laser Desorption/Ionization Time of Flight Mass Spectrometry (SELDI-TOF-MS): a potentially powerful tool for rapid characterisation of microorganisms. In Encyclopedia of Rapid Microbiological Methods Miller, M.J., ed. (DHI Publishing, LLC, River Grove, IL), vol. 3, pp. 57–96.

Shah, H.N., and Gharbia, S.E. (2011). A century of systematics of the genus *Bacteroides*: from a single genus up to the 1980s to an explosion of assemblages and the dawn of MALDI-TOF-Mass Spectrometry. The Bulletin of BISMiS 2(2), 87–106.

Shah, H.N., Keys, C., Gharbia, S.E., Ralphson, K., Trundle, F., Brookhouse, I., and Claydon, M. (2000). The application of MALDI-TOF mass spectrometry to profile the surface of intact bacterial cells. Micro. Ecol. Health Dis. *12*, 241–246.

Shah, H.N., Keys, C.J., Schmid, O., and Gharbia, S.E. (2002). Matrix-Assisted Laser Desorption/Ionisation Time of Flight Mass Spectrometry and Proteomics; a new era in anaerobic microbiology. Clin. Infect. Dis. 35, 58–64.

Matrix Assisted Laser Desorption Ionization Time-of-flight Mass Spectrometry for the Clinical Laboratory

1

Christopher D. Doern, Robert C. Jerris and
Mark D. Gonzalez

Abstract

In the last decade, few technologies have had a greater impact on clinical microbiology than matrix assisted laser desorption ionization time-of-flight mass spectrometry (MALDI-TOF MS). To say it has revolutionized the discipline would be an understatement. The ability to cost-effectively and rapidly identify microorganisms from broth and plated media by MALDI-TOF MS is replacing the more arduous, and time-consuming biochemical and antigen-based identification methods as well as some genetic sequence-based methodologies. In this chapter we succinctly review the role of MALDI-TOF MS in the routine laboratory, focusing on the key functions in pre-analytical, analytical and post-analytical perspectives. We acknowledge the thousands of papers and investigators that have contributed to our knowledge on this subject.

Introduction

The clinical microbiology laboratory is a dynamic, ever-changing entity. Providing cost-effective, rapid, accurate, clinically meaningful results to affect optimum patient care is the primary goal of every laboratory. In the USA, regulatory pressure on quality, medical necessity and reimbursement are pervasive issues of prime consideration for the contemporary laboratory. To this end, the laboratory must adapt and respond to the set of 'best practices'.

The traditional methodology for microorganism identification has been to perform a Gram stain on initial specimens, culture microorganisms on agar or in broth media, assess the significance of growth and identify appropriate microorganisms, and to perform antimicrobial susceptibility testing to direct therapy. Changes in laboratory methodologies have evolved over decades to improve all aspects of the older processes. Nucleic acid-based techniques like fluorescence *in situ* hybridization (FISH) and multiplexed real-time molecular assays have become common place in the clinical laboratory and have met the challenge of increased accuracy and speed but are limited to unique microorganism targets and often lack the ability to detect polymicrobic infections. Laboratorians must be mindful of the associated increased costs of these technologies in comparison with the traditional methods and be prepared to justify their use based on a variety of parameters, including patient outcome studies. Many different formats require additional space and in some cases specialized environments to decrease possible contamination from amplified products (Gadsby *et al.*,

2010). Continued technological advances, such as genetic sequence analysis, have furthered the move towards accurate identification of microorganisms and have become the standard for identification of microorganism groups that are difficult to identify or require prolonged time to identify by routine phenotypic methods (Clarridge, 2004). While extremely accurate, these methods suffer from prolonged turn-around time, increased expense and therefore are limited to relatively few institutions.

With the entry of MALDI-TOF MS, after culture growth, microorganisms can be identified in a matter of minutes. The bioinformatics and discriminatory ability of MALDI-TOF MS to identify a wide variety of microorganisms have set a new standard in clinical microbiological identification (Doern, 2013; Patel, 2013; Seng *et al.*, 2009).

Preanalytical

Cost justification

The purchase of a MALDI-TOF MS instrument represents one of the most expensive capital investments for a clinical microbiology laboratory. Therefore, two common approaches have been used to cost-justify the MALDI-TOF MS instruments. The first is a direct comparison to biochemical tests and the second is assessment of outcomes. These key studies have frequently been cited to justify acquisition of MALDI-TOF MS for routine use in the clinical laboratory. Several examples follow.

Gaillot and team detailed cost savings from their switch from Vitek2/API (bioMérieux) to MALDI-TOF MS (Gaillot *et al.*, 2011). Assessable data were compared over identical 1-year periods before and after implementation. A total of 33,320 and 38,624 isolates were tested in the pre versus post MALDI-TOF MS period, respectively. The costs were over $193,000 for conventional tests versus slightly over $21,000 in the MALDI-TOF MS period, which included the need to supplement MALDI-TOF MS with biochemical tests at a cost of $5374. The average cost per identification was calculated at $5.81 per isolate in the pre versus $0.41 in the post MALDI-TOF MS period. Further, they detailed decreased waste from 1424 kg to 44 kg, decreased subcultures amounting to a savings of $1102, and a decreased need for sequence analysis amounting to $1650. In total, they noted an 89% decrease in cost of identification and an annual savings of over $177,000.

Tan and co-workers conducted a performance review comparative study in routine clinical practice with a specimen based, bench to bench comparison (Tan *et al.*, 2012). For 824 bacteria and 128 yeast from 2214 specimens, the MALDI-TOF MS provided identification 1.45 days earlier and reduced labour and reagent costs by over $102,000 in 12 months. These numbers and this analysis can be customized to individual laboratories to assess cost savings and return on investment.

Similar analyses have been reported for specific specimen types, namely for specimens from cystic fibrosis (CF) patients. The isolates recovered from these specimens are among the most difficult to identify by biochemical assays and often require DNA sequence analysis for definitive identity. Desai and colleagues analysed 464 isolates from 24 unique CF specimens and detailed 85% of complex microorganisms (non-fermentative Gram-negative rods including 29 *Burkholderia* species) identified within 48 hours compared with only 34% by conventional methods (Desai *et al.*, 2012). The cost per identification using lean six sigma hand motion analysis was $1.25 by MALDI-TOF MS compared to $25.00 by conventional methods for these microorganisms.

Using MALDI-TOF MS to calculate patient outcomes and corresponding institutional savings is another approach for justifying this new technology.

Perez and co-workers applied MALDI-TOF MS identification for early diagnosis of Gram-negative bacteraemia (Perez *et al.*, 2014). They examined the impact on appropriate therapy, improvement in patient care outcomes, and total healthcare expenditures in preintervention versus intervention time periods with over 100 patients in each group. They showed identification and susceptibility results were available 22.7 hours earlier (47.1 vs. 24.4 hours), an improved time to optimal therapy of 54 hours (75 vs. 29 hours), a mean length of stay decrease of over 2 days (11.9 vs. 9.3 days), and a decrease in total hospital costs of over $19,000 ($45,709 vs. $26,162).

Nagel and colleagues additionally evaluated the impact of real-time identification of a specific pathogen group, i.e. coagulase-negative staphylococci from blood cultures by MALDI-TOF MS on antimicrobial stewardship (Nagel *et al.*, 2014). Using two time periods, pre and post MALDI-TOF MS, they showed that patients with bacteraemia in the post group were initiated on optimal therapy sooner (58.7 vs. 34.4 hours; $P = 0.030$), resulting in decreased mortality (21.7% vs. 3.1%; $P = 0.023$), and had a decreased duration of unnecessary antibiotic therapy (3.89 vs. 1.31 days; $P = 0.032$) and a decreased number of vancomycin through assays performed (1.95 vs. 0.88; $P < 0.001$).

Selection of instrument

At present there are three instruments available to perform MALDI-TOF MS: Bruker Daltonik, microflex Biotyper (Billerica, MA, USA), bioMérieux Vitek MS (Durham, NC, USA) and the Andromas SAS (Paris, France). The former two instruments are available in the USA and received clearance from the Food and Drug Administration (FDA) for certain groups of microorganisms and will be the focus of this review. The two systems are generally equivalent in cost, but differ in several ways.

The Bruker system is a table-top instrument and much smaller than the stand-alone Vitek MS. The larger size of the Vitek MS is partially due to the longer vacuum/flight tube. Specifications are detailed below.

Vitek MS (originally, Axima from Shimadzu)

Installation requirements

- Electrical – 200 VAC, 50/60 Hz, 1000 VA single phase OR 230 VAC, 50/60 Hz, 1000 VA single phase.
- UPS required to supply stable and continuous power for reliable operation.
- Temperature – ambient 18–26°C.
- Relative humidity – less than 70% non-condensing.
- Vibration free, firm, level floor, at least 330 kg supported at four points.

Laser

- 337 nm nitrogen laser, fixed focus.
- 3 ns pulse rate – 50 Hz (50 laser shots per second).

- Near normal (on-axis) incidence of the laser beam to the sample.
- Laser power and laser aim under software control.

Analyser
- Linear flight tube of 1.2 m drift length.
- Vacuum maintained by two turbomolecular pumps (nominal 250 l/s) with rotary backing.
- Beam blanking to deflect unwanted high intensity signals, e.g. matrix ions.

Mass range
- 1 kDa to 500 kDa.

Dimensions

Size:
- W 27.5 in/700 mm
- D 33.5 in/850 mm
- H 75.5 in/1920 mm

Weight:
- 727 lbs (330 kg) excluding data system.

Bruker MALDI biotyper (Microflex MS)

Installation requirements
- Electrical – 110 VAC (North America); 230 VAC (Europe); 240 VAC (Australia).
- UPS required to supply stable and continuous power for reliable operation.
- Temperature – ambient 10–30°C.
- Relative humidity – less than 18–85% non-condensing.
- Vibration free, firm, level table top.

Laser
- Nitrogen cartridge laser MNL106, 337 nm; min shots 6×10^7.
- Laser power and laser aim under software control.

Analyser
- Microflex LT.

Mass range
- 2 kDa to 200 kDa standard.

Dimensions

Size:
- W 20.1 in/510 mm.
- D 26.8 in/680 mm.
- H 43 in/1092 mm.

Weight:

- 185 lbs (83.9 kg).

Of importance for workflow is the throughput of each system. The Vitek MS can run four target plates simultaneously (i.e. 4 × 48 wells, total of 192) versus up to 96 wells on the Bruker system. In the USA, the Bruker standard is the 48-well target.

Note that there is a difference in pump time between the systems, taking longer for the Vitek MS because of it's larger loading area and longer flight tube. Because of the differences in potential sample size, it is important to assess volume and plan accordingly. Many labs that evaluate cultures by specimen type, place targets at each bench, and technologists identify microorganisms concurrently while plate reading. Another approach is to have a batch mode where a single technologist is responsible for running all specimens. Both processes have advantages and disadvantages. Further consideration must be given to workflow when lab automation and/or when the instrument is interfaced with an antimicrobial susceptibility testing system (see below).

The target plates for each instrument also differ. The Vitek MS relies solely on disposable targets, while the Bruker offers both a reusable polished steel target as well as disposable targets. Third party manufacturers are now also offering disposable target for both systems (uFocus, Hudson Surface Technology, Ft. Lee, NJ, USA). These alternatives must be validated prior to use and at present are not allowed in IVD workflow.

The key points here are:

- Steel targets are reusable. They require routine cleaning with ethanol and also a thorough cleaning at least monthly requiring a hazardous organic material (trifluoroacetic acid, used under a fume hood). Appropriate disposal of organic solvents is also required. While disposable targets add to the cost, users cite an advantage in workflow and avoidance of chemicals.
- While service contracts are expensive, it is essential for machine maintenance. The instruments are relatively simple in nature, but vacuum pumps, mother-boards, and lasers are all subject to wear and tear and break down. Further, routine instrument maintenance is required for optimal performance. In this regard, several procedural things can be done to maintain top performance. First, always keep a target in the instrument under vacuum when not in use. Second, make sure to run standards on a per-run basis. These should include the respective system supplied standard(s), biological standards with specifically designated microorganisms (see 'CAP requirements, below) and matrix only (to ensure no 'memory' proteins that may contaminate both matrix and that may be present on the Bruker reusable steel target). In addition, it is critical to maintain electronic communication with the respective technical services to tweak the instruments as needed. Therefore, preanalytical planning must include data drops for remote diagnostics.

Culture condition environment (aerobic, anaerobic, microaerophilic, microaerobic), temperature, and media have little effect on accuracy of identification of microorganisms by MALDI-TOF MS. In general, non-selective media yields better identification than selective media (Anderson *et al.*, 2012). Lower definitive identification rates with staphylococci were noted with colistin-nalidixic acid agar versus Columbia sheep blood agar plates (75% vs.

95%, respectively). Of significant note from their study was that sampling from selective media was not associated with incorrect identifications. The study had a small sample size and lacked statistical support for their conclusions.

Additionally, microorganisms can be tested after storage. McElvania and colleagues detailed the ability to accurately identify a subset of microorganisms after 5 days of storage at 35°C and 4°C (McElvania Tekippe *et al.*, 2013). Our experience extends this to 7–10 days post routine incubation. However, in our hands, microorganisms generally do not give identifications when tested from refrigerated plates (personal experience).

Pre-analytical considerations will change based also on automation that may accompany the instruments. There are tracking platforms and totally automated systems to assist with seeding the microorganisms on the target. Available for the Bruker:

- Bruker Galaxy
- Bruker Pilot
- Pickolo-MI, Robotic sampling, TECAN (Morrisville, NC, USA)
- Copan WASP (Murrietta, CA, USA)
- Copan, MALDI-Trace (Murrietta, CA, USA)
- Becton Dickinson (BD) Kiestra (Franklin Lakes, NJ, USA).

These systems facilitate paperless, guided target preparation through one of several mechanisms including upfront data entry into a traceable log (e.g. BD Kiestra), barcoded sample entry (e.g. Bruker Pilot) or radiofrequency identification (RCID) (e.g. Copan MALDI Trace).

The bioMérieux Vitek MS linked to the Vitek 2 antimicrobial susceptibility testing (AST) has built-in software to facilitate tracking and seeding of MALDI-TOF MS plates.

Consideration should be given to integration with AST. As previously noted, the Vitek MS can be integrated into the Vitek 2 AST. To date the Bruker MALDI Biotyper has the capacity to be integrated into at least three ASTs: the BD Phoenix (Franklin Lakes, NJ, USA), the TREK Aris (THERMO Scientific, Oakwood Village, OH, USA) and MicroScan (Beckman Coulter, Brea, CA, USA).

Regulatory FDA (USA)

In the USA, these instruments fall under the purview of the FDA for regulatory clearance. Both the Vitek MS (Vitek MS IVD) and the Bruker MALDI Biotyper (MALDI Biotyper CA, USA) have been granted clearance for select microorganisms. The Vitek MS was granted 510(k) clearance in August 2013 for 192 microorganisms including yeast (*Candida, Cryptococcus,* and *Malassezia* groups), and bacteria from the *Staphylococcaceae, Streptococcaceae, Enterobacteriaceae, Pseudomonadaceae* and *Bacteroidaceae* families.

The Bruker MALDI Biotyper system was approved in November of 2013 with a claim for identification of 100 Gram-negative species or species groups. FDA clearance in April of 2015 heralded an additional 170 species and species groups, representing 180 clinically relevant species of aerobic Gram-positives, fastidious Gram-negatives, *Enterobacteriaceae,* anaerobic bacteria and yeasts (http://ir.bruker.com/investors/press-releases/press-release-details/2015/Bruker-Announces-FDA-Clearance-for-Second-Expanded-Claim-for-the-MALDI-Biotyper-CA-System/default.aspx, accessed 15 October 2015).

Verification and validation of MALDI-TOF MS performance

Before a MALDI-TOF MS system can be used for clinical diagnostics, laboratories must verify (FDA-cleared devices) or validate (laboratory developed tests (LDT)) their instruments and databases. These are studies designed to assess the performance of a specific instrument in a given laboratory. The benefit of utilizing an FDA-approved device is that laboratories can have confidence that the performance of the instrument has been extensively evaluated through rigorous clinical trials. As a result the extent of the testing required to verify performance in an individual laboratory is minimal and well-defined. Specific guidance for the verification of FDA-cleared MALDI-TOF MS instruments is in process and will be published by the Clinical and Laboratory Standards Institute.

For LDTs it is less clear what is required to validate the instrument. Currently, MALDI-TOF MS is a completely new technique that has not been previously applied to microorganism identification. Therefore, the design of validation studies is at the discretion of the laboratory director. In a survey of several major academic medical centre laboratories using LDT MALDI-TOF MS there was some amount of consensus in what was done to validate their instruments (Doern, 2013). Generally speaking, laboratories validated their LDT MALDI-TOF MS systems by testing around 1000 total isolates and comparing the MALDI identifications to conventional laboratory methods and resolving discrepancies with 16S rRNA sequencing. All laboratories elected to perform their analyses by microorganism category. That is to say, Gram-negatives, Gram-positives, and yeasts were all analysed separately.

A few other components of testing were also validated by these laboratories. Different types of media were evaluated (i.e. blood agar vs. chocolate agar vs. MacConkey, etc.) as well as incubation conditions (i.e. duration and temperature) and spot method (i.e. toothpick vs. swab, etc.) (Doern, 2013). Several studies have looked at the impact of these various conditions on microorganism identification. Anderson and colleagues looked at impact of various solid media on microorganism identification with MALDI-TOF MS and found that overall the method was capable of identifying microorganisms from a variety of different media (Anderson *et al.*, 2012). In addition, MALDI-TOF MS can identify microorganisms from liquid media such as that which is used to culture mycobacteria or blood cultures (Buchan *et al.*, 2014; Mediavilla-Gradolph *et al.*, 2015). Identification from positive blood cultures and patient specimens is discussed below.

A unique feature of the MALDI-TOF MS databases is that they can readily be updated to improve the quality of system performance. These updates include the addition of new spectra as well as the removal of poor or incorrect spectra. There is no guidance as to how laboratories should validate their database updates and few laboratories have attempted to do so. It would seem unnecessary to completely revalidate the entire database by testing 1000+ isolates. Rather, laboratories may consider testing a targeted subset of isolates that would best evaluate the updated database. This could be done by taking isolates all the way through the analysis process, from culture to identification. Alternatively, laboratories may choose to reanalyse the spectra that was captured during their initial validation study. The benefit of this approach is that it allows a direct comparison in score values by eliminating the variable of a biological replicate test.

Note that current interface capabilities with the Bruker instrument are not uniform in the ability to use self-validated or self developed databases. Users are encouraged to communicate directly with manufacturers for specifications.

CAP requirements

The College of American Pathologists recognizes MALDI-TOF MS as a valid method to identify microorganisms. To that end, they have included in their latest checklist for US Laboratory Accreditation, a list of mandatory requirements for compliance.

These include:

1. Instrument operation: Procedures are documented for operation and calibration of the mass spectrometer.
2. A calibration control is run each day of patient/client testing, with each change in target plate, or according to manufacturer's recommendations and these records are maintained.
3. Appropriate control microorganisms are tested on a daily basis to include at least one bacterium, a representative yeast, other microorganism (*Mycobacterium, Nocardia*) if included in the run [subjected to the same extraction procedures as tested isolates], and a blank control. Records are maintained.
4. Reagents and solvents are of appropriate grade (HPLC grade or other). MSDS sheets are maintained with reagent logs.
5. Consumables are of appropriate manufacturing type to function as required.

Analytical

Test performance

A critical component of identifying a microorganism with MALDI-TOF MS is the template or target plate onto which the microorganisms are applied. These target plates are made of a polished steel surface with designated areas where microorganisms are to be applied. There are two primary types of targets plates that can be used for microorganism identification: reusable and disposable.

As previously noted, reusable target plates may save cost over time but require routine cleaning with 70% ethanol, mass spectrometer grade water, and 80% trifluoroacetic acid. For safety, the cleaning process should be performed under a chemical fume hood while wearing two pairs of nitrile gloves. There are several soak steps involved in cleaning and the entire process can take up to 30 min. Cleaning need not be performed after each run if there are unused wells available. Users can simply perform additional runs by spotting microorganisms to unused wells. One concern may be that the repeated use of targets without cleaning may expose laboratory staff and the MALDI-TOF MS to potentially infectious material. However, the process of adding the alpha-cyano-4-hydroxycinnamic acid (CHCA) is thought to be rapidly bactericidal and probably eliminates any risk of transmitting infectious material from re-used target plates.

Aside from the target plates shown in Figs. 1.1 and 1.2, which are commonly used for microorganism identification, there are additional targets that can be used to serve other purposes. For example, there are targets with a smaller number of larger wells which can be used when it is not possible to fit a specimen into the smaller well of the standard target. In addition, there are 'anchor plates' which have hydrophobic areas surrounding a small central location which collects and concentrates the specimen. These specialized targets can be used to analyse dilute liquid specimens.

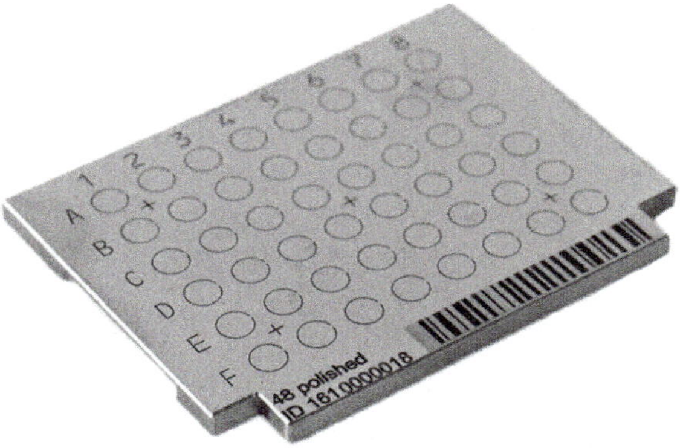

Figure 1.1 Bruker steel, reusable target.

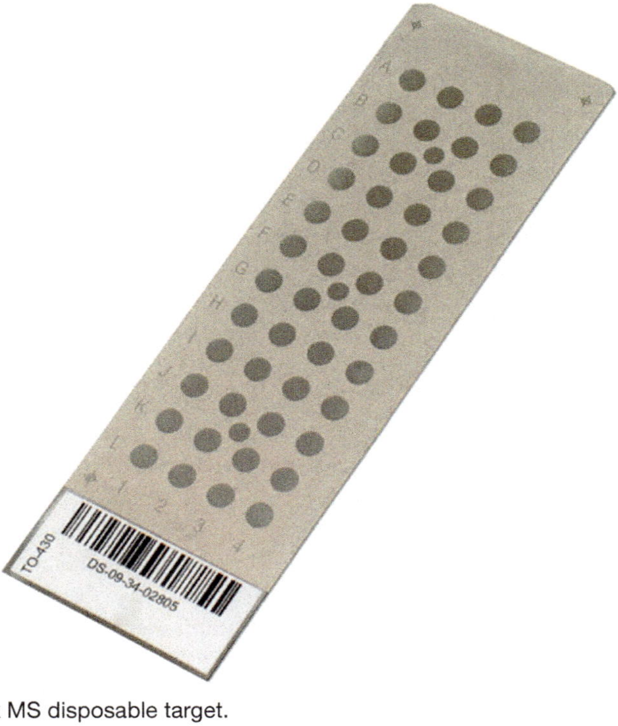

Figure 1.2 Vitek MS disposable target.

Preparation

There are a variety of methods that can be used to prepare a microorganism for MALDI-TOF MS analysis. These methods include, direct microorganism application, direct formic acid overlay, and protein extraction. The process of extracting protein for analysis and microorganism identification varies by microorganism type and will be discussed accordingly.

The simplest method for performing microorganism identification is direct spotting of the microorganism, allowing it to dry and overlaying the microorganism with 1 µl of CHCA and allowing that to dry. CHCA is prepared in a volatile solution that dries within 3–5 minutes. This process has been shown to produce reliable identifications for both Gram-positive and Gram-negative bacteria (Tan *et al.*, 2012). Although some have found that additional steps are required for the reliable identification of Gram-positive bacteria (Alatoom *et al.*, 2011). In addition, mucoid or highly encapsulated microorganisms can be more difficult to identify with the standard matrix-only method (Marko *et al.*, 2012).

A supplementary method that can be used to enhance the quality of spectra obtained from microorganisms is the on-plate extraction or formic acid overlay method. Using this method, microorganisms are spotted directly to the plate and then 1 µl of formic acid (70–100%) is applied to the microorganism and allowed to dry. This step helps to extract proteins so that they can be more readily available for identification. Once the formic acid is dry, the CHCA matrix is added and the MALDI-TOF MS is performed as per normal protocols. This method has been shown to improve the chances of obtaining quality spectra and results in higher overall identification scores (Bessède *et al.*, 2011). There is little evidence that this method improves the identification of Gram-negative bacteria, but some have shown that it improves the identification of Gram-positive bacteria and it appears to be critical for the identification of yeast (Alatoom *et al.*, 2011; McElvania Tekippe *et al.*, 2013).

The two methods discussed above provide a simple workflow; however, they are occasionally unable to generate spectra that results in microorganism identification. For those difficult to identify microorganisms, a full extraction or in-tube extraction can be performed. The disadvantage of this process is that it requires the use of harsh chemicals and includes several centrifugation steps that are more labour intensive than the on-plate methods. While manufacturer recommendations may vary slightly, the general extraction methods involve the transfer of microorganisms into a tube containing mass spectrometry grade water followed by thorough vortexing. An aliquot of 100% ethanol is then added to a final concentration of about 75% and is followed by a second vortex step. The sample is then pelleted through centrifugation, the supernatant is removed, and then 70% formic acid is added followed by an equal amount of 100% acetonitrile and a vortex step. The solution is then centrifuged, and the supernatant can then be used for MALDI-TOF MS analysis.

Although yeast can be identified using the relatively simple formic acid overlay method, filamentous fungi require additional processing. Currently, there are no FDA-cleared databases for the analysis of filamentous fungi and as a result a variety of methods have been used to process the microorganisms for identification. An exhaustive discussion of all published methods is outside the scope of this chapter so what follows is a general description of the methods. The literature clearly demonstrates that it is critical that extraction is required for efficient identification of filamentous fungi. In addition, these can be dangerous microorganisms with spores that readily aerosolize. As a result, safety is of primary concern when developing a MALDI-TOF MS extraction method for filamentous fungi. Inactivation of mould is required prior to analysis and several methods have been explored for this purpose. They include ethanol treatment, heat treatment, and mechanical lysis. The growth condition is also a critical factor that impacts mould identification (i.e. broth vs. agar growth) (personal experience, RCJ). Regardless of method used, it seems clear that reliable identification will only be possible if the extraction method used for analysis is the same as the one used to generate the database spectra (Lau *et al.*, 2013).

As with filamentous fungi, special processing methods must be employed for the identification of *Nocardia* and mycobacteria and many of the same principles apply. Verroken *et al.* (2010) described what appears to be a robust method for extracting and identifying *Nocardia*. This method calls for the suspension of 10 colonies in water which are then boiled and centrifuged. Ethanol is then added to the supernatant to precipitate the proteins. These proteins are collected through centrifugation and resuspended in 70% formic acid and acetonitrile and then analysed by MALDI-TOF MS. With respect to mycobacterial identification, no consensus methods have been developed. Safety is of paramount concern when identifying isolates that may be *Mycobacterium tuberculosis* so although formic acid processing alone has yielded successful identifications, there are significant safety issues with this method. Subsequently, a number of studies have been published and describe varying inactivation and protein purification steps for the identification of mycobacteria (Lotz *et al.*, 2010; Saleeb *et al.*, 2011). These methods rely on either extended heating steps or mechanical disruption for the inactivation of the microorganism. Viability experiments have shown both techniques to be effective in killing the mycobacteria and ensuring safety (Wilen *et al.*, 2015). It is advised that all labs perform their own analysis to optimize safety.

Spotting technique

In protocols where microorganisms are spotted directly to a target plate from an isolated colony there are several ways in which this can be accomplished. It is important when applying a microorganism, that it be distributed evenly on the target plate to minimize three dimensional structures. Three-dimensional structure in the specimen can have a deleterious effect on the quality of spectra captured and result in unsuccessful identifications. Anecdotally, there appears to be a significant learning curve for technologists as they learn to apply microorganisms to the target plate. This is surprising given how simple the procedure appears to be at face value. However, optimal microorganism spotting requires smooth application of specimen as well as the correct amount of microorganism. There is an element of subjectivity in deciding how much microorganism to apply and success rates invariably improve with experience. Many laboratories are using MALDI-TOF MS as a laboratory developed test (LDT) and are using non-standardized application methods. Common techniques include the use of toothpicks, swabs, pipette tips and plastic loops. All of these can be used successfully but require some practice.

Matrix

Matrix is a critical component of mass analysis and the type of analysis being performed dictates what matrix should be used. There are many matrices that can be used and they have profound effects on sensitivity and accuracy of the method. Cinnamic acid is the basis for most matrices and that is the case for the most commonly used matrix in whole-cell microorganism identification, alpha-cyano-4-hydrocinnamic acid (CHCA). A review of the application of other matrices is outside the scope of this chapter but suffice to say, there are numerous options that can be selected for different applications of MALDI-TOF MS. It is important to note that matrix assisted laser desorption ionization is commonly done from solid samples. Therefore, it is important that the matrix solution be allowed to completely evaporate after it is deposited on the specimen. This allows the solvent to be removed, leaving a solid sample that is embedded in matrix.

Workflow

MALDI-TOF MS is a method unlike any other in the clinical microbiology laboratory. As a result, laboratories have an opportunity to rethink their workflow so that they can take advantage of improved turn-around times offered by MALDI-TOF MS. Laboratory organization and workflow varies from institution to institution and is dictated by the patient population served, test volume, operating hours, amongst many other factors. As laboratories consider implementing MALDI-TOF MS into their workflow they will need to consider several key factors that are discussed below.

Laboratory Information System (LIS) and instrument interface

Most clinically relevant microorganisms that are definitively identified in the laboratory will have antimicrobial susceptibility testing (AST) performed. Prior to implementation of MALDI-TOF MS most laboratories will have relied on automated, growth-based systems for both identification and AST. As a result, information transfer between the identification and AST were seamless and not an issue. With the implementation of MALDI-TOF MS, laboratories will have to evaluate about how best to integrate the identification with the AST. In many cases, AST systems require that a microorganism name be provided prior to releasing a susceptibility result to the LIS. It then becomes important to consider how the MALDI-TOF MS will be interfaced. Will the instrument be able to interface directly with the AST system or will it communicate with the LIS? Depending on the regulatory status of the MALDI-TOF MS system being used, laboratories may not be able to interface with either and will have to manually enter their information into the LIS or AST system.

Centralized versus decentralized MALDI-TOF MS processing

There are numerous nuances to be considered when designing laboratory workflow around MALDI-TOF MS. Generally speaking though, these can be broken down into two main categories, centralized and decentralized processing. For the purposes of this discussion, these terms are used to refer to testing performed within a single laboratory, not in terms of reference testing.

The centralized approach to processing MALDI-TOF MS will rely on a single individual for the processing of target plates. This structure will be very similar to way many laboratories already process AST, which is that plates are brought to a dedicated person who is responsible for setting up and performing AST. In MALDI-TOF MS workflows, laboratories may elect to have technologists bring agar plates to a dedicated MALDI person who will perform microorganism identification. In this set-up, laboratories must have protocols in place that allow the bench technologist to indicate which isolates are to be tested. This can easily be accomplished as long as there are well-isolated colonies that can be circled with a marker or wax pencil. Alternatively, laboratories can distribute target plates to each bench and have the technologists spot their own targets. Once these targets are spotted, they can be brought to a dedicated individual who will pipette the reagents and process the targets. This work may be advantageous in avoiding errors in selecting microorganisms for identification, but may be disadvantageous in that it will slow down the bench technologist. It also requires that a larger number of technologists be trained to spot target plates and maintain competency.

The decentralized approach dictates that each bench technologist be responsible for the entire MALDI-TOF process. As discussed above, this approach requires that a large number of technologists be trained and competent in performing MALDI-TOF MS, which may be a deterrent for laboratories with a large number of bench technologists. This approach might also lead to bottle necks in processing target plates as MALDI-TOF MS instruments are not designed to be random access. However, some manufacturers are able to process multiple target plates at one time and may better accommodate such a workflow.

Quality control

The necessary quality control for MALDI-TOF MS differs substantially from that of other methods. At the outset, laboratories were performing MALDI-TOF MS as laboratory developed tests and had to develop their own quality control measures. As with other identification systems, laboratories typically included a biological positive control. Some would even include a rotating schedule of microorganisms that would be used to ensure that the system was performing adequately for all microorganism types. The most common systems being used for microorganism identification also mandate that a standard be used to calibrate the instrument. For the Bruker Biotyper, this standard includes the spectra of *Escherichia coli* DH5 alpha along with two additional proteins that extend the upper mass range that can be calibrated. This standard not only calibrates the instrument but it also serves as another level of quality control as the instrument should be able to produce an identification of *E. coli* from the standard. For non-FDA cleared devices, the frequency of calibration was at the discretion of the laboratory director. Some laboratories elected to calibrate with every run while others would calibrate less frequently. The Vitek MS uses direct testing from a designated ATCC strain of *E. coli*.

While most laboratories included positive controls, some also included negative reagent controls. These are controls in which blank reagents are added and the wells are assessed to ensure that no contaminated peaks are present. This control helps to ensure that reagents have not been contaminated and for those labs re-using target plates, it ensures that the cleaning process has been successful. It takes ~10^5 colony-forming units (CFU) to generate an identification with MALDI-TOF MS (Croxatto *et al.*, 2012), so it is unlikely that a reagent or target plate would become so completely contaminated that it would generate a false identification. However, contaminating protein generates additional spectra that can lower your identification scores and reduce the chances of successfully identifying a microorganism.

Direct from specimen identification

MALDI-TOF MS rapidly and reliably identifies microorganisms from cultured media. The speed and accuracy of the method has led microbiologists to explore its use in direct-from-specimen testing. As stated above, MALDI-TOF MS requires approximately 10^5 CFU to generate an identification, which renders it unable to produce identifications directly from most specimens. However, there are a few scenarios in which MALDI-TOF MS can work directly from a specimen. Most commonly, MALDI-TOF MS is used to identify microorganisms that grow from positive blood cultures. Due to the high microorganism burden in these bottles they make for acceptable specimens for MALDI-TOF MS identification. The only other specimen that has been attempted with any success thus far is urine. The methods and performance for both of these specimens will be discussed briefly below.

Identification from positive blood culture bottles

Currently, there are no FDA cleared MALDI-TOF MS based assays for direct identification of microorganisms from positive blood cultures. (Note that in 2015, an IVD-CE Sepsityper protocol has been approved for use outside the USA.) As a result, there is no consensus method but there are some commonalities between methods that are found in the peer-reviewed literature. Subtle differences in protocol account for unique characteristics of the MALDI-TOF MS instrument as well as the blood culture media being used. Generally, these protocols involve the following steps:

1. A lysis buffer is added to the blood culture specimen, which is then vortexed and centrifuged.
2. The pellet is washed (in some cases multiple times) with a wash buffer and re-centrifuged if necessary.
3. The pellet is suspended in ethanol and centrifuged.
4. The pellet is suspended in an acetonitrile and formic acid solution and centrifuged.
5. The supernatant is then analysed on the MALDI-TOF MS.

Solution concentrations, centrifuge steps, and wash steps vary by protocol, but most studies in the peer-reviewed literature follow a variation of this protocol.

A complicating factor in these analyses is the presence of charcoal in blood culture media as it has been shown to have a deleterious effect on the frequency of identification (Fiori *et al.*, 2014; Fothergill *et al.*, 2013). Some have experimented with alternative protocols that involve the filtering of blood culture solutions followed by a scraping step that obtains the microorganism for analysis (Machen *et al.*, 2014).

Despite method variation, numerous studies have now shown that microorganisms can successfully be identified from positive blood cultures regardless of instrument and blood culture media. The actual success rates for individual microorganism categories varies by study but it does appear that MALDI-TOF MS systems perform more reliably for the identification of Gram-negative bacteria than for Gram-positive bacteria (Buchan *et al.*, 2012; Chen *et al.*, 2013; Fothergill *et al.*, 2013). Yeast can also be identified from blood culture bottles but the success rates achieved in the literature vary widely (Buchan *et al.*, 2012; Fothergill *et al.*, 2013; Schubert *et al.*, 2011). In addition, mixed cultures pose some problems as MALDI-TOF MS often fails to produce any identification in these circumstances, and when an identification is produced, only one microorganism will be identified (Buchan *et al.*, 2012; Chen *et al.*, 2013; Fothergill *et al.*, 2013).

Although, protocols for identifying microorganisms directly out of positive blood cultures are more labour intensive than identifying microorganisms cultured on solid media, the improvement in turn-around time have significantly improved patient care. Several studies have now shown that when combined with a robust stewardship effort, MALDI-TOF MS can decrease hospital costs and improve outcomes (Perez *et al.*, 2014). Finally, there is a commercially available purification system, Sepsityper (Bruker Daltonics, Billerica, MA, USA) that can be used for cleanup of positive blood cultures prior to MALDI-TOF MS analysis.

Identification from urine

Given the 10^5 CFU limit of detection, urine seems like a logical specimen to attempt to perform direct MALDI-TOF MS identifications. Many laboratories use a quantitative cut-off of 10^5 CFU/ml for defining positivity, which in combination with the fact that most urinary tract infections (UTI) are monomicrobic, naturally suggests that MALDI-TOF MS can be used to diagnose UTI.

A handful of studies have evaluated MALDI-TOF MS in this capacity. One of the first studies by Ferriera *et al.* (2011) used a differential centrifugation method to remove leucocytes followed by a high-speed centrifugation to concentrate the bacteria. In their study they found that 92.7% and 91.8% of specimens that grew $\geq 10^5$ CFU yielded a genus and species-level identification, respectively. Additionally, they reported that adding a formic acid and acetonitrile step significantly improved their results.

DeMarco and Burnham (2014) later published their own protocol that involved a soft spin centrifugation (1000*g*) followed by a diafiltration step. Remaining red blood cells (RBCs) were lysed with the addition of Milli-Q water. The specimen was then desalted and concentrated with several wash steps. Lastly, the pellet from the last wash step was pipetted onto a MALDI-TOF MS target plate and analysed. This study found that Gram-negative UTI could be correctly classified for *E. coli*, *Klebsiella pneumoniae* and *Proteus mirabilis* using this method. In addition, negative specimens were also correctly classified. In the final analysis, this method achieved 100% specificity but only 67% sensitivity.

Currently, direct identification from urine specimens remains an unproven method and has not become common practice for the diagnosis of UTI. More research will be required to improve the workflow and reliability of this method.

Postanalytical

No system for identification is without problems and the same is true with MALDI-TOF MS. To cite a few of the most common issues: the inability to separate certain microorganisms (based on genotypic/protein profile similarity); the absence of microorganisms from the database; the identification of an apparently unusual microorganism that is closely related to a commonly known microorganism (that has been identified by phenotypic methods), and; identification of a microorganism that has not been previously recognized. It is of significant note that erroneous identifications are extremely rare.

To elucidate, *Shigella* and *E. coli* are so closely related that *Shigella* is not identified by MALDI-TOF MS. To differentiate, colonies should be examined for lactose fermentation, motility and a spot indole. If positive an identification of *E. coli* is appropriate. If negative for these tests, *Shigella* should be suspected and an agglutination with *Shigella* antisera performed. If there are issues with agglutination, a phenotypic assay can readily differentiate the species of *Shigella*. Progress towards differentiating these two genera by MALDI-TOF MS are progressing (Khot and Fisher, 2013; Paauw *et al.*, 2015).

Using the original database analysis, *Streptococcus pneumoniae* could not be distinguished from *S. mitis*. As such colony morphology, optochin disc susceptibility (pneumococcus is susceptible), bile solubility (pneumococcus is soluble) or a pneumococcus specific agglutination assay (pneumococcus is positive) can be used to differentiate the two bacteria. It is particularly important to identify these bacteria correctly as *S. mitis/oralis* frequently represents non-pathogenic colonization while *S. pneumoniae* is a pathogen capable of causing serious and life-threatening disease. Fortunately, with the newest improved databases (and full extraction with the Bruker system) there is now evidence that both MALDI-TOF MS systems are capable of differentiating these bacteria, which should obviate the need to rely on optochin disk and bile solubility testing (Chen *et al.*, 2015; Manji *et al.*, 2014).

The ability to definitively identify organisms to the genus and species level that are unfamiliar to clinicians and laboratorians alike, poses a particular problem. Using the Bruker system, we have encountered a number of these in our experiences and have chosen to lump these into a 'group' or 'complex' of microorganisms with a designation familiar to practitioners. Several of these are listed in Table 1.1.

The microbiologist must work closely with their information technology team to accommodate these changes in both the AST and LIS interfaces. Commercial manufacturers are also working closely with the MS systems to optimize interfaces.

Microorganisms that are identified that have not been previously described also pose an issue for reporting. If not previously validated, the microorganism must be identified by sequence analysis or alternative reference methods. If a microorganism has been validated, communication directly with the clinician is warranted. A description of the organisms can be found in a number of online databases including http://www.bacterio.net/, http://www.ncbi.nlm.nih.gov/guide/taxonomy/, http://www.dsmz.de/ or http://www.bacterio.cict.fr.

It is important to note that many of these organisms have not been validated for interpretive criteria for susceptibility testing. As such, it is recommended to simply give MIC values.

For Vitek MS a reading of 60.0% to 99.9% with a single identification is criterion to accept identification (typically to species level). The same limitations as noted above apply to the Vitek MS. Since this instrument is traditionally interfaced with the Vitek II AST, most of the identification/reporting is seamless. When a result from the Vitek MS is not in the Clinically Applicable (CA) database, the instrument will flag the result, not send the data to the AST and requires the user to manually input identification into the AST.

For the Bruker system, Table 1.2 denotes the log score and interpretation. Note that many investigators have validated scores that amend the above for a number of organism groups.

Postanalytical

Post-analytical evaluation of MALDI-TOF MS organism identification

An important feature of MALDI-TOF MS performance is that it rarely produces incorrect identifications when confidence scores are high. As result, laboratories can confidently report MALDI-TOF MS microorganism identifications with minimal confirmatory testing. However, supplemental biochemical testing can still play an important role in assessing MALDI-TOF MS results as there are several situations that can lead to erroneous microorganism identification. These situations include clerical error, erroneous microorganism selection, erroneous target spotting, and analysis errors made by the MALDI-TOF MS system itself. The following is a discussion about how best to recognize and prevent these errors to ensure optimal post-analytical performance.

Table 1.1 Examples of microorganism names and groups encountered in MALDI-TOF MS identification

Species/group/ complex	Strains included in database	Different species are potentially associated with the displayed identification. Based on 16S rRNA gene sequencing a secure species differentiation between the displayed species is difficult. Confirmatory tests are required to differentiate between listed microorganisms
Achromobacter xylosoxidans	*A. xylosoxidans*	*A. xylosoxidans*, *A. denitrificans*, *A. insolitus*, *A. marplatensis*, *A. ruhlandii*, *A. spanius*
Acinetobacter baumannii complex	*A. baumannii*, *A. calcoaceticus*, *A. pittii*, *A. nosocomialis*	*A. baumannii*, *A. calcoaceticus*, *A. pittii*, *A. nosocomialis*
Aeromonas spp.	*A. allosaccharophila*, *A. caviae*, *A. culicicola*, *A. hydrophila*, *A. ichthiosmia*, *A. sobria*, *A. veronii*	*A. allosaccharophila*, *A. aquariorum*, *A. caviae*, *A. culicicola*, *A. enteropelogenes*, *A. fluvialis*, *A. hydrophila*, *A. ichthiosmia*, *A. jandaei*, *A. media*, *A. punctata*, *A. rivuli*, *A. sanarellii*, *A. sobria*, *A. taiwanensis*, *A. veronii*
Burkholderia gladioli	*B. gladioli*	*B. gladioli*, *B. glumae*, *B. caryopylii*
Burkholderia multivorans	*B. multivorans*	*B. multivorans*
Burkholderia cepacia complex	*B. ambifaria*, *B. anthina*, *B. cenocepacia*, *B. cepacia*, *B. diffusa*, *B. dolosa*, *B. lata*, *B. latens*, *B. metallica*, *B. pyrrocinia*, *B. seminalis*, *B. stabilis*, *B. vietnamiensis*	*B. ambifaria*, *B. anthina*, *B. cenocepacia*, *B. cepacia*, *B. diffusa*, *B. dolosa*, *B. lata*, *B. latens*, *B. metallica*, *B. pyrrocinia*, *B. seminalis*, *B. stabilis*, *B. vietnamiensis*
Citrobacter amalonaticus complex	*C. amalonaticus*, *C. farmeri*	*C. amalonaticus*, *C. farmeri*
Citrobacter freundii complex	*C. braakii*, *C. freundii*, *C. gillenii*, *C. murliniae*, *C. rodentium*, *C. sedlakii*, *C. werkmannii*, *C. youngae*	*C. braakii*, *C. freundii*, *C. gillenii*, *C. murliniae*, *C. rodentium*, *C. sedlakii*, *C. werkmannii*, *C. youngae*
Enterobacter cloacae complex	*E. asburiae*, *E. cancerogenus*, *E. cloacae*, *E. hormaechei*, *E. kobei*, *E. ludwigii*	*E. asburiae*, *E. cancerogenus*, *E. cloacae*, *E. cowanii*, *E. hormaechei*, *E. kobei*, *E. ludwigii*, *E. mori*, *E. nimipressuralis*, *E. soli*
Escherichia coli	*E. coli*	*E. albertii*, *E. coli*, *E. fergusonii*, *Shigella* spp.
Haemophilus influenzae	*H. influenzae*	*H. aegyptius*, *H. influenzae*
Hafnia alvei	*Hafnia alvei*	*H. alvei*, *H. paralvei*, *Obesumbacterium proteus*
Klebsiella pneumoniae	*Klebsiella pneumoniae*	*K. pneumoniae*, *K. granulomatis*, *K. singaporensis*, *K. variicola*
Klebsiella oxytoca	*K. oxytoca*	*K. oxytoca*, *R. ornithinolytica*
Raoultella ornithinolytica	*R. ornithinolytica*	*R. planticola*

Table 1.1 Continued

Species/group/ complex	Strains included in database	Different species are potentially associated with the displayed identification. Based on 16S rRNA gene sequencing a secure species differentiation between the displayed species is difficult. Confirmatory tests are required to differentiate between listed microorganisms
Moraxella osloensis	*M. osloensis*	*Enhydrobacter aerosaccus*, *Moraxella osloensis*
Morganella morganii	*Morganella morganii*	*M. morganii*, *M. psychrotolerans*
Pantoea agglomerans	*Pantoea agglomerans*	*P. agglomerans*, *P. anthophila*, *P. brenneri*, *P. conspicua*, *P. eucalypti*, *P. vagans*
Proteus vulgaris group	*P. hauseri*, *P. penneri*, *P. vulgaris*	*P. hauseri*, *P. penneri*, *P. vulgaris*
Providencia rettgeri	*Providencia rettgeri*	*P. rettgeri*, *P. alcalifaciens*, *P. burhodogranariea*, *P. heimbachae*, *P. rustigianii*, *P. vermicola*
Pseudomonas aeruginosa	*P. aeruginosa*	
Pseudomonas fluorescens group	*P. congelans*, *P. corrugata*, *P. extremorientalis*, *P. fluorescens*, *P. gessardii*, *P. libanensis*, *P. mandelii*, *P. marginalis*, *P. migulae*, *P. mucidolens*, *P. orientalis*, *P. poae*, *P. rhodesiae*	*P. congelans*, *P. corrugata*, *P. extremorientalis*, *P. fluorescens*, *P. gessardii*, *P. libanensis*, *P. mandelii*, *P. marginalis*, *P. migulae*, *P. mucidolens*, *P. orientalis*, *P. poae*, *P. rhodesiae*, *P. synxantha*, *P. tolaasii*, *P. trivialis*, *P. veronii*
Stenotrophomonas maltophilia	*S. maltophilia*, *Pseudomonas beteli*, *Ps. hibiscola*, *Ps. geniculata*	*S. maltophilia*, *Pseudomonas beteli*, *Pseudomonas hibiscola*, *Pseudomonas geniculata*
Yersinia pseudotuberculosis	*Y. pseudotuberculosis*	*Y. pestis*, *Y. pseudotuberculosis*, *Y. similis*

Table 1.2 Interpretation of Bruker log(score) values

Range	Interpretation	Colour
2.00–3.00	High confidence identification	Green
1.70–1.99	Low confidence identification If this log(score) is obtained on a direct transfer, follow with extraction preparation (see User Manual) If this log(score) is obtained on a extracted test organism, report sample as 'low confidence identification'	Yellow
<1.70	No organism identification possible (Refer to the troubleshooting section in User Manual) If this log(score) is obtained on a direct transfer, follow with extraction preparation (see User Manual) If this log(score) is obtained on a extracted test organism, report sample as 'no identification'	Red

Performance limitations

Regardless of the system being used, MALDI-TOF MS has proven to be a robust and reliable method for identifying even the most challenging microorganisms (Deak *et al.*, 2015; McElvania TeKippe and Burnham, 2014; Pence *et al.*, 2014; Wilen *et al.*, 2015). However, several microorganism categories are readily identified by mass spectrometry and laboratories must be wary of these limitations when reporting identifications. Such limitations include differentiation of *E. coli* from *Shigella* spp. as well as differentiation of *S. pneumoniae* from the *S. mitis/oralis* group (see above).

Limitations in MALDI-TOF MS identification of *S. pneumoniae, S. mitis/oralis, E. coli* and *Shigella* spp. are well-described. However, MALDI-TOF MS can make errors in some unpredictable ways and thus it is critical that clinical microbiologists maintain their knowledge of basic biochemical reactions, Gram stain morphology, and colony morphology. This point is best illustrated by Alby *et al.* (2015), who reported a series of misidentification by MALDI-TOF MS but could have been caught with a basic knowledge of microorganism characteristics. MALDI-TOF MS has proven to be an extremely reliable method, but this study demonstrates that post-analytical evaluation of identifications remains an important step in microorganism identification.

FDA-cleared versus laboratory-validated database

Post-analytical analysis may vary depending on whether a laboratory is using an FDA-cleared database or a laboratiry-validated database. Both present distinct advantages and disadvantages that need to be considered.

FDA cleared databases for *in vitro* diagnostics (IVD) are valuable because their performance is extensively evaluated through multicentre clinical trials. Owing to the rigours and expense of an FDA clinical trial, these IVD databases are somewhat limited in their microorganism coverage. The two currently available FDA cleared MALDI-TOF MS databases contain approximately 200 microorganisms. While this is enough to identify the vast majority of clinical isolates, they lack some rare, but clinically important species. As a result, laboratories may face situations in which they have to resort to alternative identification methods, such as 16S rRNA gene sequencing.

Conversely, laboratory validated databases can contain a much larger number of microorganisms. However, for the same reason that FDA cleared databases fail to claim these rare microorganisms, laboratories may struggle to test enough isolates in their validation studies to assess performance. As such, a laboratory that encounters an unvalidated microorganism should have post-analytical protocols in place to ensure that the identification is correct. Some form of confirmatory testing such as a unique biochemical reaction, colony morphology, or ideally 16S rRNA gene sequencing should be performed prior to reporting the result.

In addition, strain representation is a critical factor that impacts the ability of MALDI-TOF MS to reliably identify an isolate. In some circumstances laboratories may encounter microorganisms that are represented in the databases but are not readily identified due to poor strain representation. A post-analytic process that can improve identification in such a case is the addition of that microorganism's spectra to a database. In this way, a laboratory can develop their own, institution specific, database that will improve the performance of their system (Marklein *et al.*, 2009). We stress that, while potentially beneficial, the addition of new spectra to a database is a complex process that requires great attention to detail and an understanding of how to generate quality spectra. More importantly, laboratories need to ensure that they have correctly identified the microorganism that is being added to the database. Full gene 16S rRNA sequencing should be used in lieu of phenotypic identification techniques to confirm microorganism identity prior to adding it to a database for clinical diagnostics. Obviously this is outside of the routine IVD and should be coordinated with the manufacturers.

Lastly, MALDI-TOF MS has been a transformative technology for the clinical microbiology laboratory that facilitates the accurate identification of microorganisms that labs were previously unable to identify on a routine basis. As such, laboratories are now able to report unusual microorganisms with which providers may be unfamiliar. When unusual identifications are encountered, laboratories should consider first whether the isolate is clinically significant. The reporting of unusual, but clinically insignificant organisms will probably lead to confusion and overtreatment. If an organism is deemed to be clinically significant, laboratories may want to consider including an interpretive comment to assist providers in interpreting the result.

MALDI-TOF MS laboratory information system interfaces

Once MALDI-TOF MS analysis has completed, results must be transferred to the laboratory information system (LIS) for reporting. With the wide-spread adoption of MALDI-TOF MS laboratory validated systems, many users are faced with transferring results manually because FDA cleared AST and LISs were not compatible with existing MALDI-TOF MS software. Reporting of results in this setting requires a tedious and error-prone process by which information is manually entered into the LIS. This of course presents opportunities for clerical error and delays time to reporting.

Over time, MALDI-TOF MS manufacturers have developed relationships with the manufacturers of most commonly used AST systems, which has facilitated the development of interfaces that allow for the direct transfer of MALDI-TOF MS results to the instrument itself, middleware, or to the LIS. Bruker has relationships with Becton Dickinson, Trek, Beckman Coulter and BioMIC, which are responsible for manufacturing the BD Epicenter middleware, the Aris, the Microscan, and the BioMIC, respectively. In addition, Bruker has relationships with Sunquest, Cerner, Soft, and Epic which should facilitate direct transfer of information into those LISs. The bioMerieux Vitek MS IVD system is capable of communicating with the Vitek 2 AST system as well as the Myla middleware package. Lastly, third-party solutions are being developed that will aid in interfacing these instruments with products with which no formal relationship exists.

Summary

The acquisition and implementation of MALDI-TOF MS in the clinical microbiology laboratory requires considerable investigation of resources and time. The various pre-analytical, analytical and post-analytical factors discussed above provide a framework for bringing on MALDI-TOF MS. Nonetheless, the many advantages of MALDI-TOF MS, which includes a rapid, cost-effective, and accurate identification of a range of microorganisms, warrant the consideration of this new revolutionary technology.

References

Alatoom, A.A., Cunningham, S.A., Ihde, S.M., Mandrekar, J., and Patel, R. (2011). Comparison of direct colony method versus extraction method for identification of gram-positive cocci by use of Bruker Biotyper matrix-assisted laser desorption ionization-time of flight mass spectrometry. J. Clin. Microbiol. *49*, 2868–2873.

Alby, K., Glaser, L.J., and Edelstein, P.H. (2015). *Kocuria rhizophila* misidentified as *Corynebacterium jeikeium* and other errors caused by the Vitek MS system call for maintained microbiological competence in the era of matrix-assisted laser desorption ionization-time of flight mass spectrometry. J. Clin. Microbiol. *53*, 360–361.

Anderson, N.W., Buchan, B.W., Riebe, K.M., Parsons, L.N., Gnacinski, S., and Ledeboer, N.A. (2012). Effects of solid-medium type on routine identification of bacterial isolates by use of matrix-assisted laser desorption ionization-time of flight mass spectrometry. J. Clin. Microbiol. *50*, 1008–1013.

Bessède, E., Angla-Gre, M., Delagarde, Y., Sep Hieng, S., Ménard, A., and Mégraud, F. (2011). Matrix-assisted laser-desorption/ionization biotyper: experience in the routine of a University hospital. Clin. Microbiol. Infect. *17*, 533–538.

Buchan, B.W., Riebe, K.M., and Ledeboer, N.A. (2012). Comparison of the MALDI Biotyper system using Sepsityper specimen processing to routine microbiological methods for identification of bacteria from positive blood culture bottles. J. Clin. Microbiol. *50*, 346–352.

Buchan, B.W., Riebe, K.M., Timke, M., Kostrzewa, M., and Ledeboer, N.A. (2014). Comparison of MALDI-TOF MS with HPLC and nucleic acid sequencing for the identification of Mycobacterium species in cultures using solid medium and broth. Am. J. Clin. Pathol. *141*, 25–34.

Chen, J.H., Ho, P.L., Kwan, G.S., She, K.K., Siu, G.K., Cheng, V.C., Yuen, K.Y., and Yam, W.C. (2013). Direct bacterial identification in positive blood cultures by use of two commercial matrix-assisted laser desorption ionization-time of flight mass spectrometry systems. J. Clin. Microbiol. *51*, 1733–1739.

Chen, Y., Porter, V., Mubareka, S., Kotowich, L., and Simor, A.E. (2015). Rapid identification of bacteria directly from positive blood cultures using a serum separator tube, smudge plate preparation, and Matrix-Assisted Laser Desorption Ionization-Time of Flight Mass Spectrometry (MALDI-TOF MS). J. Clin. Microbiol. *50*, 3349–3352.

Clarridge, J.E. (2004). Impact of 16S rRNA gene sequence analysis for identification of bacteria on clinical microbiology and infectious diseases. Clin. Microbiol. Rev. *17*, 840–862.

Croxatto, A., Prod'hom, G., and Greub, G. (2012). Applications of MALDI-TOF mass spectrometry in clinical diagnostic microbiology. FEMS Microbiol. Rev. *36*, 380–407.

Deak, E., Charlton, C.L., Bobenchik, A.M., Miller, S.A., Pollett, S., McHardy, I.H., Wu, M.T., and Garner, O.B. (2015). Comparison of the Vitek MS and Bruker Microflex LT MALDI-TOF MS platforms for routine identification of commonly isolated bacteria and yeast in the clinical microbiology laboratory. Diagn. Microbiol. Infect. Dis. *81*, 27–33.

Demarco, M.L., and Burnham, C.A. (2014). Diafiltration MALDI-TOF mass spectrometry method for culture-independent detection and identification of pathogens directly from urine specimens. Am. J. Clin. Pathol. *141*, 204–212.

Desai, A.P., Stanley, T., Atuan, M., McKey, J., Lipuma, J.J., Rogers, B., and Jerris, R. (2012). Use of matrix assisted laser desorption ionisation-time of flight mass spectrometry in a paediatric clinical laboratory for identification of bacteria commonly isolated from cystic fibrosis patients. J. Clin. Pathol. *65*, 835–838.

Doern, C.D. (2013). Charting Uncharted Territory: A review of the verification and implementation process for Matrix-Assisted Laser Desorption Ionization-Time of Flight Mass Spectrometry (MALDI-TOF MS) for organism identification. Clin. Microbiol. Newsletter 35, 69–78.

Ferreira, L., Sánchez-Juanes, F., Muñoz-Bellido, J.L., and González-Buitrago, J.M. (2011). Rapid method for direct identification of bacteria in urine and blood culture samples by matrix-assisted laser desorption ionization time-of-flight mass spectrometry: intact cell vs. extraction method. Clin. Microbiol. Infect. *17*, 1007–1012.

Fiori, B., D'Inzeo, T., Di Florio, V., De Maio, F., De Angelis, G., Giaquinto, A., Campana, L., Tanzarella, E., Tumbarello, M., Antonelli, M., *et al.* (2014). Performance of two resin-containing blood culture media in detection of bloodstream infections and in direct matrix-assisted laser desorption ionization-time of flight mass spectrometry (MALDI-TOF MS) broth assays for isolate identification: clinical comparison of the BacT/Alert Plus and Bactec Plus systems. J. Clin. Microbiol. *52*, 3558–3567.

Fothergill, A., Kasinathan, V., Hyman, J., Walsh, J., Drake, T., and Wang, Y.F. (2013). Rapid identification of bacteria and yeasts from positive-blood-culture bottles by using a lysis-filtration method and matrix-assisted laser desorption ionization-time of flight mass spectrum analysis with the SARAMIS database. J. Clin. Microbiol. *51*, 805–809.

Gadsby, N.J., Hardie, A., Claas, E.C., and Templeton, K.E. (2010). Comparison of the Luminex Respiratory Virus Panel fast assay with in-house real-time PCR for respiratory viral infection diagnosis. J. Clin. Microbiol. *48*, 2213–2216.

Gaillot, O., Blondiaux, N., Loïez, C., Wallet, F., Lemaître, N., Herwegh, S., and Courcol, R.J. (2011). Cost-effectiveness of switch to matrix-assisted laser desorption ionization-time of flight mass spectrometry for routine bacterial identification. J. Clin. Microbiol. *49*, 4412.

Khot, P.D., and Fisher, M.A. (2013). Novel approach for differentiating Shigella species and Escherichia coli by matrix-assisted laser desorption ionization-time of flight mass spectrometry. J. Clin. Microbiol. *51*, 3711–3716.

Lau, A.F., Drake, S.K., Calhoun, L.B., Henderson, C.M., and Zelazny, A.M. (2013). Development of a clinically comprehensive database and a simple procedure for identification of molds from solid media by matrix-assisted laser desorption ionization-time of flight mass spectrometry. J. Clin. Microbiol. *51*, 828–834.

Lotz, A., Ferroni, A., Beretti, J.L., Dauphin, B., Carbonnelle, E., Guet-Revillet, H., Veziris, N., Heym, B., Jarlier, V., Gaillard, J.L., *et al.* (2010). Rapid identification of mycobacterial whole cells in solid and liquid culture media by matrix-assisted laser desorption ionization-time of flight mass spectrometry. J. Clin. Microbiol. *48*, 4481–4486.

Machen, A., Drake, T., and Wang, Y.F. (2014). Same day identification and full panel antimicrobial susceptibility testing of bacteria from positive blood culture bottles made possible by a combined lysis-filtration method with MALDI-TOF VITEK mass spectrometry and the VITEK2 system. PLoS One *9*, e87870.

Manji, R., Bythrow, M., Branda, J.A., Burnham, C.A., Ferraro, M.J., Garner, O.B., Jennemann, R., Lewinski, M.A., Mochon, A.B., Procop, G.W., *et al.* (2014). Multi-center evaluation of the VITEK® MS system for mass spectrometric identification of non-Enterobacteriaceae Gram-negative bacilli. Eur. J. Clin. Microbiol. Infect. Dis. *33*, 337–346.

Marklein, G., Josten, M., Klanke, U., Müller, E., Horré, R., Maier, T., Wenzel, T., Kostrzewa, M., Bierbaum, G., Hoerauf, A., *et al.* (2009). Matrix-assisted laser desorption ionization-time of flight mass spectrometry for fast and reliable identification of clinical yeast isolates. J. Clin. Microbiol. *47*, 2912–2917.

Marko, D.C., Saffert, R.T., Cunningham, S.A., Hyman, J., Walsh, J., Arbefeville, S., Howard, W., Pruessner, J., Safwat, N., Cockerill, F.R., *et al.* (2012). Evaluation of the Bruker Biotyper and Vitek MS matrix-assisted laser desorption ionization-time of flight mass spectrometry systems for identification of nonfermenting gram-negative bacilli isolated from cultures from cystic fibrosis patients. J. Clin. Microbiol. *50*, 2034–2039.

McElvania TeKippe, E., and Burnham, C.A. (2014). Evaluation of the Bruker Biotyper and VITEK MS MALDI-TOF MS systems for the identification of unusual and/or difficult-to-identify microorganisms isolated from clinical specimens. Eur. J. Clin. Microbiol. Infect. Dis. *33*, 2163–2171.

McElvania Tekippe, E., Shuey, S., Winkler, D.W., Butler, M.A., and Burnham, C.A. (2013). Optimizing identification of clinically relevant Gram-positive organisms by use of the Bruker Biotyper matrix-assisted laser desorption ionization-time of flight mass spectrometry system. J. Clin. Microbiol. *51*, 1421–1427.

Mediavilla-Gradolph, M.C., De Toro-Peinado, I., Bermúdez-Ruiz, M.P., García-Martínez, M.e.L., Ortega-Torres, M., Montiel Quezel-Guerraz, N., and Palop-Borrás, B. (2015). Use of MALDI-TOF MS for identification of nontuberculous mycobacterium species isolated from clinical specimens. Biomed. Res. Int. *2015*, 854078.

Nagel, J.L., Huang, A.M., Kunapuli, A., Gandhi, T.N., Washer, L.L., Lassiter, J., Patel, T., and Newton, D.W. (2014). Impact of antimicrobial stewardship intervention on coagulase-negative Staphylococcus blood cultures in conjunction with rapid diagnostic testing. J. Clin. Microbiol. *52*, 2849–2854.

Paauw, A., Jonker, D., Roeselers, G., Heng, J.M., Mars-Groenendijk, R.H., Trip, H., Molhoek, E.M., Jansen, H.J., van der Plas, J., de Jong, A.L., *et al.* (2015). Rapid and reliable discrimination between Shigella species and Escherichia coli using MALDI-TOF mass spectrometry. Int. J. Med. Microbiol. *305*, 446–452.

Patel, R. (2013). Matrix-assisted laser desorption ionization-time of flight mass spectrometry in clinical microbiology. Clin. Infect. Dis. *57*, 564–572.

Pence, M.A., McElvania TeKippe, E., Wallace, M.A., and Burnham, C.A. (2014). Comparison and optimization of two MALDI-TOF MS platforms for the identification of medically relevant yeast species. Eur. J. Clin. Microbiol. Infect. Dis. *33*, 1703–1712.

Perez, K.K., Olsen, R.J., Musick, W.L., Cernoch, P.L., Davis, J.R., Peterson, L.E., and Musser, J.M. (2014). Integrating rapid diagnostics and antimicrobial stewardship improves outcomes in patients with antibiotic-resistant Gram-negative bacteremia. J. Infect. *69*, 216–225.

Saleeb, P.G., Drake, S.K., Murray, P.R., and Zelazny, A.M. (2011). Identification of mycobacteria in solid-culture media by matrix-assisted laser desorption ionization-time of flight mass spectrometry. J. Clin. Microbiol. *49*, 1790–1794.

Schubert, S., Weinert, K., Wagner, C., Gunzl, B., Wieser, A., Maier, T., and Kostrzewa, M. (2011). Novel, improved sample preparation for rapid, direct identification from positive blood cultures using matrix-assisted laser desorption/ionization time-of-flight (MALDI-TOF) mass spectrometry. J. Mol. Diagn. *13*, 701–706.

Seng, P., Drancourt, M., Gouriet, F., La Scola, B., Fournier, P.E., Rolain, J.M., and Raoult, D. (2009). Ongoing revolution in bacteriology: routine identification of bacteria by matrix-assisted laser desorption ionization time-of-flight mass spectrometry. Clin. Infect. Dis. *49*, 543–551.

Tan, K.E., Ellis, B.C., Lee, R., Stamper, P.D., Zhang, S.X., and Carroll, K.C. (2012). Prospective evaluation of a matrix-assisted laser desorption ionization-time of flight mass spectrometry system in a hospital clinical microbiology laboratory for identification of bacteria and yeasts: a bench-by-bench study for assessing the impact on time to identification and cost-effectiveness. J. Clin. Microbiol. *50*, 3301–3308.

Verroken, A., Janssens, M., Berhin, C., Bogaerts, P., Huang, T.D., Wauters, G., and Glupczynski, Y. (2010). Evaluation of matrix-assisted laser desorption ionization-time of flight mass spectrometry for identification of *nocardia* species. J. Clin. Microbiol. *48*, 4015–4021.

Wilen, C.B., McMullen, A.R., and Burnham, C.A. (2015). Comparison of sample preparation methods, instrumentation platforms, and contemporary commercial databases for identification of clinically relevant *mycobacteria* by Matrix-Assisted Laser Desorption Ionization-Time of Flight Mass Spectrometry. J. Clin. Microbiol. 53, 2308–2315.

Analysis of Anaerobes and Some Other Fastidious Bacteria

2

Elisabeth Nagy

Abstract

Anaerobes, like some other fastidious, slow-growing bacteria belonging to the HACEK group have special place among human pathogenic bacteria. They are important pathogens, however, owing to their special requirements to be isolated from clinical samples, such as strict anaerobic or CO_2-rich environment for their long incubation time, makes their identification time-consuming. In this chapter several applications of the mass spectrometry will be discussed beside the rapid identification of these bacteria. It has been shown since the first studies that the database developments are mandatory including not only reference strains of the different well-known or newly accepted species, but including the mass spectra of well-characterized clinical isolates may improve the performance of the MALDI-TOF MS identification of these bacteria. Mass spectrometry-based typing of *Bacteroides fragilis* strains may help to detect clinical isolates belonging to division II, which harbour resistance gene against carbapenems. *Propionibacterium acnes* phylotypes could also be distinguished by looking for peak variations of the mass spectra of the isolates belonging to this species. Direct identification of anaerobic bacteria from positive blood cultures should be evaluated further as well as the possibilities to use MALDI-TOF MS for antibiotic resistance determination of these bacteria.

Introduction

The indigenous bacterial flora on mucosal membranes of humans and animals is dominated by bacteria, which can not multiply and form colonies on the surface of solid media in the presence of oxygen (Finegold and George, 1989). Several hundreds of different anaerobic species have been identified during the past decades by time-consuming classical and DNA-base molecular methods. It has been proven that many of them can cause severe, life-threatening infections alone or being part of a mixture of aerobic and anaerobic pathogens. The majority of anaerobic bacteria need a long incubation time in the anaerobic environment due to slow growth to get enough biomass to use classical, automated or miniaturized identification methods that are based on selected biochemical tests. However, all these methods do not satisfy clinicians and help in timely patient care. MALDI-TOF MS revolutionized the identification of anaerobic and other slow-growing, fastidious bacteria obtained in pure culture or being part of a mixed infection.

Early developments of databases for anaerobes

The early studies using mass spectrometry in the field of anaerobic bacteria showed that species belonging to the three frequently isolated Gram-negative genera *Bacteroides, Porphyromonas* and *Prevotella* can easily be distinguished by the comparison of their MS profiles over the range of 0.5–3 kDa. Surface-enhanced laser desorption/ionization time-of-flight mass spectrometry (SELDI-TOF-MS) was used, which is a modification of MALDI-TOF MS and applies a ProteinChip array instead of the target plate used for MALDI-TOF MS (Shah *et al.*, 2002). Furthermore, some representative members of the genus *Porphyromonas* (*Porphyromonas maccacae, Porphyromonas catoniae, Porphyromonas canoris* and *Porphyromonas endodontalis*), which are normally difficult to differentiate by conventional methods, furnished consistent and highly characteristic MS patterns. These early studies also showed that two phylogenetically closely related *Prevotella* species, *Prevotella intermedia* and *Prevotella nigrescens*, which can not be separated by biochemical tests or gas–liquid chromatography could be distinguished by MALDI-TOF MS using mathematical similarity analysis (Shah *et al.*, 2002; Stingu *et al.*, 2008). Mass spectrometry analysis also showed that *Prevotella nigrescens* is a more homogeneous species in terms of its surface molecules, whereas *Prevotella intermedia* may be divided in different subgroups by this method (Shah *et al.*, 2002). During classification of a great variety of *Clostridium* spp. it was possible to achieve the clear separation of *Clostridium chauvoei* and *Clostridium septicum* by mass spectrometry, two clinically important species, which are normally difficult to differentiate by traditional methods. During the next years using MALDI-TOF MS for the identification of anaerobes, it became evident that the development of databases of the different systems is mandatory if routine laboratories want to use this method for identification of clinically relevant anaerobic bacteria. These database developments were done by routine laboratories, specialized for culturing clinically relevant anaerobes in collaboration with one or other company, or was carried out in their own laboratory by incorporating in the database new, well-characterized species and confirming its applicability by testing earlier not identified isolates with success (Grosse-Herrenthey *et al.*, 2008; Nagy *et al.*, 2009; Veloo *et al.*, 2011a; Wybo *et al.*, 2012; Stingu *et al.*, 2015). Several clinically important genera of anaerobic bacteria were involved in these database developments.

Different species of *Clostridium* are among the most important anaerobic bacteria causing severe infections. Some species can be identified easily, but some others may cause problems to be identified (Jousimies-Somer *et al.*, 2002). A study involving 64 *Clostridium* isolates belonging to 31 species (most of them originating from different type culture collections) was dedicated to early development of the Biotyper database of Bruker (Germany) (Grosse-Herrenthey *et al.*, 2008). Further, 25 clinical and environmental *Clostridium* isolates, identified by biochemical tests and sequencing, were added to prove the discriminatory power of MALDI-TOF MS for this genus. During this study, the effects of different culture media as well as the prolonged incubation time in the anaerobic environment on the MS fingerprints of clostridia were also tested. It was found that only the advanced state of sporulation was unfavourable for the exact identification of clostridia. Recent publication (Chean *et al.*, 2014) compared two systems (Vitek MS and Bruker MS) for the identification of clinically relevant *Clostridium* isolates, evaluating also the impact of the sample preparation and the completeness of the databases. They found that out of 52 blood culture isolates belonging to 10 *Clostridium* spp., identified by sequencing, Vitek MS identified 47 (90.4%) isolates on species level by the 'direct transfer' sample preparation method, whereas

Bruker MS needed the 'extended direct transfer method' adding formic acid to the sample on the target plate to identify all 52 (100%) isolates. In another study with the Bruker MS, only 1 of 66 *Clostridium* isolates, representing 12 species, gave an unacceptable identification with a log score 1.406, which proved to be *Clostridium hathewayi* by sequencing, and was not included in the Biotyper database at that time (Nagy *et al.*, 2012).

Strains belonging to the *Bacteroides*/*Parabacteroides* genus are among the most frequently isolated anaerobic bacteria in routine clinical microbiology laboratories. Owing to limited database in the classical identification kits several clinically relevant species such as *Bacteroides cellulosilyticus, Bacteroides nordii, Bacteroides dorei* or newly accepted species such as *Bacteroides clarus, Bacteroides fluxus* and *Bacteroides oleiciplenus* (Watanabe *et al.*, 2010) are not included. During a Europe-wide antibiotic resistance surveillance, where species level differences in the resistance was looked for, 277 clinical isolates, belonging to the genera *Bacteroides* and *Parabacteroides*, were tested by MALDI-TOF MS. Using the 'full formic acid extraction' for sample preparation 97.5% of the 277 strains were correctly identified on species level (log score ≥ 2.0) by the Bruker MS showing already a rather developed database for these anaerobes (Nagy *et al.*, 2009). If the phenotypic identification differed from the mass spectrometry result 16S rRNA gene sequencing was used which confirmed the MALDI-TOF MS results in all cases. For all 7 isolates, which gave inconclusive results with Bruker MS (log score < 1.7), the sequence data revealed that they were missing from the Biotyper database at that time. After the inclusion of a reference spectrum of one of the four *Parabacteriodes distasonis* isolates in the database, the three others were identified with high log scores (> 2.5). The mass spectra of three further *Bacteroides* species (*Bacteroides eggerthii, Bacteroides goldsteinii* and *Bacteroides intestinalis*) were included into the Biotyper database as a consequence of this investigation (Nagy *et al.*, 2009).

During an extensive study published by Wybo *et al.* (2012) about the species identification of clinical *Prevotella* isolates by MALDI-TOF MS, the results were compared with 16S rRNA gene sequencing. Through use of the Bruker MS system (Reference Library 3.2.1.0), which included the spectra of 20 *Prevotella* species, 62.7% of 102 *Prevotella* clinical isolates were identified at the species level, and 73.5% at the genus level. The commercial database was extended in-house with the spectra of 23 further *Prevotella* reference strains (adding 13 different species which were missing from the database), and this improved the species- and genus-level identification of the 102 clinical isolates from 62.7% to 83.3% and from 73.5% to 89.2%, respectively. Following the addition of a sequenced clinical isolate of *Prevotella heparinolytica* to the database, the identification of two other *Prevotella heparinolytica* isolates of the collection was possible by MALDI-TOF MS.

Gram positive anaerobic cocci (GPACs, including 13 genera and at least 33 species at the present time) are difficult to identify by conventional methods due to their inactivity in biochemical tests. Recent taxonomic changes make their identification even more difficult. Using the AXIMA (Shimadzu) MALDI-TOF MS equipment, 12 sequenced reference strains belonging to six genera and 12 different species as well as 77 sequenced clinical isolates were applied (Veloo *et al.*, 2011a). Species specific identifying spectra, so called SuperSpectra, were constructed and added to the SARAMIS database. The performance of the constructed database was challenged by recent isolates and 90% of the 107 unknown clinical isolates of GPACs were correctly identified. Three of 32 *Finegoldia magna*, one of two *Anaerococcus vaginalis* and two of three *Peptoniphilus ivorii* were not identified when 'direct transfer' sample preparation was applied. One *Peptoniphilus octavius*, which was not included in the database at that time, and three GPACs, which could not be given a name by 16S rRNA gene sequencing, were not identified either.

An extensive evaluation of the Biotyper (Bruker) database for *Actinomyces* species was done by Stingu *et al.*, (2015). Mass spectra obtained from 11 reference strains and 140 sequenced clinical isolates were used to create the reference database. A cross-validation of this reference database representing 14 *Actinomyces* species yielded correct identification for all species, which were represented by more than two strains in the database. For *Actinomyces naeslundii* and *Actinomyces johnsonii* as well as *Actinomyces meyeri* and *Actinomyces odontolyticus*, similar to sequencing data, the mass spectra were not discriminatory.

Performance of the mass spectrometer-based identification of clinically relevant anaerobes in routine laboratories

Between 2011 and 2014 several studies were designed to utilize MALDI-TOF MS (MALDI Biotyper or Vitek MS) for the identification of anaerobic bacteria in routine clinical microbiological laboratories and compare the MS-related identification results with those of classical, phenotypic identification methods (La Scola *et al.*, 2011; Federko *et al.*, 2012; Culebras *et al.*, 2012; Knoester *et al.*, 2012; Nagy *et al.*, 2012; Coltella *et al.*, 2013; Kierzkowska *et al.*, 2013; Barreau *et al.*, 2013; Garner *et al.*, 2014; Barba *et al.*, 2014; Lee *et al.*, 2015). In the event of discrepant results, 16S rRNA gene sequencing in most cases confirmed the MALDI-TOF MS identification. In a few studies 16S rRNA gene sequencing was used as a 'gold standard' for the identification of isolated anaerobes if MALDI-TOF MS failed to give a species-level identification (Justesen *et al.*, 2011; Barreau *et al.*, 2013). Some studies compared the performance of the available two systems and databases for the identification of anaerobic bacteria (Veloo *et al.*, 2011b; Justesen *et al.*, 2011; Martiny *et al.*, 2012; Jamal *et al.*, 2013). These studies differ in the number of anaerobic isolates tested (< 100 to > 1000), as well as in the number and composition of the genera (between 2 and 32) and the species (between 13 and 102) involved, mainly dependent on the clinical background of the laboratories (Nagy, 2014). During the routine use of the mass spectrometry-based identification for anaerobes in different laboratories several rare and/or recently described anaerobic species were identified by the Biotyper, such as *Anaerotruncus colihominis, Anaerococcus murdochii, Anaerococcus tetradius, Dialister micraerophilus, Porphyromonas gulae, Bacteroides heparinolyticus, Bacteroides salyersiae, Bacteroides tecticus, Prevotella nanceiensis, Prevotella baroniae* or *Turicibacter sanguinis* (Barreau *et al.*, 2013) which shows the applicability of this method for the routine identification of a great variety of the clinically important anaerobic bacteria. The performance of the MS-based identification of anaerobic bacteria is very much influenced by the database available at the time of the study as well as identification criteria (cut-off values for species or genus level identification) provided by the manufacturers or selected by the laboratories. The percentage of the correct species-level identification of strict anaerobic bacteria experienced during these different studies varied between 70.8% and 93.8%, and the genus-level identification varied between 88% and 98%. The most divers anaerobic species (102 different species belonging to 39 genera) were tested by Schmidt *et al.* (2013) including less frequent anaerobic isolates, which may explain the relatively low level of correct species identification (70.8%) using the cut off ≥ 2.0 log score during this study. Some groups based on great experience in anaerobe culture techniques accept lower log scores (≥1.7 to ≥1.9) for species level identification of anaerobic bacteria (La Scola *et al.*, 2011; Barreau *et al.*, 2013; Schmitt *et al.*, 2013) by this significantly increasing the percentage of the acceptable, correct identification of anaerobic species. Two database versions of the Bruker system were evaluated in the Anaerobes Reference Unit, Public Health Wales, Cardiff, UK on a wide range of anaerobes from clinical sources referred to them for species determination by 16S rRNA gene sequencing as the reference laboratory (Copsey *et al.*, 2013). Altogether 1195 isolates, representing 200 different anaerobic species and 60 anaerobic genera (21 Gram-negative and 39 Gram-positive genera), were identified by mass spectrometry using the direct smear sample preparation method on the target plate in duplicate. The older database contained 4111 entries and identified 63% of the isolates on genus or species level. This was increased using the updated version of the database containing 4613 entries, when 71% of the isolates were correctly identified at least on genus level. Many rare isolates of Gram-negative genera (*Anaerospirillum, Desulfovibrio, Selenomonas, Sneathia, Sutterella*) and Gram-positive genera (*Abiotrophia, Anaerostipes, Catabacter, Collinsella, Caprobacillus, Flavonifractor, Moryella, Parascardovia, Propioniferax, Solobacterium, Tissirella, Turicibacter*) were represented only with one isolate in both database versions and some of these seemed not be included in both database versions. The best identification was found for *Bacteroides* (81 isolates representing 11 species), *Clostridium* (130 isolates representing 28 species), *Propionibacterium acnes* (51 isolates) and *Finegoldia magna* (17 isolates) by the updated database with species identification of 91%, 88%, 71% and 94% of the isolates, respectively.

Despite the fact that fewer papers are evaluating the Vitek MS system and its IVD database for species identification of anaerobic bacteria than the Bruker MS system, there are no significant differences in the performance of the two systems for the common clinical isolates of anaerobic species (Justesen *et al.*, 2011; Jamal *et al.*, 2013; Martiny *et al.*, 2012; Garner *et al.*, 2013; Lee, *et al.*, 2015).

The biodiversity of some anaerobic species is well known. It has been proven during these studies that the expansion of the reference spectra libraries included to the different MS systems is important to be continued to optimize the performance of the mass spectrometry systems for correct identification of anaerobic bacteria. Including well characterized (16S rDNA/protein gene sequenced) clinical isolates may improve the performance. In those cases where some species such as *Anaerococcus vaginalis, Campylobacter (Bacteroides) urealyticus, Finegoldia magna* or *Clostridium hathewayi* were represented in the Biotyper database only by one or two entries, they obviously did not cover the natural variability of these species (Veloo *et al.*, 2011b; Nagy *et al.*, 2012). Out of the frequently isolated genera clinical isolates of *Fusobacterium* and *Actinomyces* are still difficult to differentiate on species level by MALDI-TOF MS, which may be explained by the wide heterogeneity of some species (such as *Fusobacterium nucleatum*) due to potential horizontal gene transfer (Claypool *et al.*, 2010). To use different sample preparation methods may also be important to improve species level identification of some anaerobic isolates. These studies show exactly how the development of the MS databases may influence the quality of species identification in the case of anaerobes.

Despite the extensive extension of the databases, there are still examples, where both MALDI-TOF MS systems failed to yield species-level identification, such as in the case of *B. vulgatus* and the closely related, newly described *Bacteroides dorei* or *Bacteroides ovatus* and *Bacteroides xylanisolvent*. The Shimadzu/SARAMIS system gave a double result (*Bacteroides vulgatus/Bacteroides dorei* and *Bacteroides ovatus/Bacteroides xylanisolvent*), whereas the

Biotyper (Bruker) misidentified these species (Justesen *et al.*, 2011). Even careful comparison of the mass spectra of these species pairs just not separates them showing the possible limitation of this technology.

Pre-analytical requirements for species level identification of anaerobic bacteria

Several pre-analytical requirements, such as the composition of the media, the incubation time, the sample preparation, the exposure to oxygen before sample preparation, were tested by several authors. It has been shown that the sporulation of clostridia may display deviations in the fingerprint patterns. The more advanced the state of sporulation is when the measurement is carried out, the greater the difference in the spectra can be observed (Grosse-Herrenthey *et al.*, 2008). A recent study clearly demonstrated that even selective media used in the routine laboratory procedure for isolation of anaerobic bacteria, as well as the number of subcultures of the isolates do not influence correct identification if 'extended direct transfer method' (on-target extraction with formic acid) is used (Hsu *et al.*, 2014). All the selected 28 anaerobic isolates, representing the most frequent clinically relevant anaerobes were correctly identified independent on which media they were cultured. The prolonged incubation time and the exposure to ambient air before sample preparation up to five days did not influence correct identification (Hsu *et al.*, 2014; Veloo *et al.*, 2014).

Different sample preparation methods may effect correct species level identification of anaerobic bacteria. The reference spectra for anaerobic bacteria in Bruker MS system were included after full chemical extraction by formic acid and acetonitrile to break the cell wall and expose intracellular proteins. This procedure requires more hands-on time as compared with use of the direct smear of the cultured colony or the on-target extraction by formic acid before MS measurement. Fournier *et al.* (2012) compared two pre-analytical processes for 238 anaerobic isolates representing 34 species. Although direct smear sample preparation led to significantly lower log score values for clostridia and GPACs, the identification of species of both genera did not differ. However, altogether fewer isolates were identified following direct smear sample preparation than after extraction (207 vs. 218 of 238 tested isolates, respectively). On the other hand, the direct smear sample preparation resulted in acceptable species identification for nine of the 20 isolates, which were not identified after the extraction (*Parvimonas micra, Anaerococcus hydrogenalis, Clostridium clostridioforme, Clostridium hathewayi, Clostridium ramosum* and four *Propionibacterium acnes*), and this was confirmed by 16S rRNA gene sequencing. The level of identification of *Bacteroides, Fusobacterium* and *Veillonella* isolates was not significantly influenced by the sample preparation, whereas for Gram-positive anaerobes with more rigid cell wall, such as *Clostridium* spp. and GPACs full chemical extraction gave better species level identification with higher log scores (Furnier *et al.*, 2012). Another study indicated that the identification of anaerobic bacteria and especially that of *Clostridium ramosum* could be improved, if the second run was processed by 70% formic acid pre-treatment on the target plate (Justesen *et al.*, 2011). Some other Gram-positive anaerobic species such as *Bifidobacterium longum*, and *Finegoldia magna* could be identified better if the on-target extraction was used, however *Peptoniphilus ivorii* needed a full extraction for species level identification (Veloo *et al.*, 2014). No different sample preparation method, other than the direct smear on the target plate, is recommended or even allowed for anaerobic bacteria before measurement in the Vitek MS system (Garner *et al.*, 2013).

Use of MALDI-TOF MS for the typing of anaerobic bacteria

It has been shown during the past years that MALDI-TOF MS can also distinguish some bacteria belonging to the same species. This may provide a tool for typing of certain bacteria for epidemiological or other purposes. Protein-based typing by MALDI-TOF MS measurement was carried out first on *Mycobacterium tuberculosis* (Hettick *et al.*, 2006) *Listeria monocytogenes* (Barbuddhe *et al.*, 2008), *Streptococcus agalactiae* (Lartigue *et al.*, 2009), *Lactococcus lactis* (Tanigawa, *et al.*, 2010) and methicillin resistant *Staphylococcus aureus* (Wolters *et al.*, 2011) and the results proved to be comparable with those of classical or molecular typing methods such as MLST, *spa* typing or PFGE.

There are few clinically important anaerobes where sub-species level typing is important for epidemiological or other reasons. *Clostridium difficile* causing severe diarrhoea in the hospital settings as well as in the community became one of those species where typing is extremely important to follow the spread of more virulent or antibiotic resistant subtypes. The most frequent typing method for *Clostridium difficile* is PCR-ribotyping, which is usually performed in specialized laboratories and on the other hand it is time-consuming and expensive. Using a standard collection of 25 different *Clostridium difficile* PCR ribotypes, a database was set up from the mass spectra of these strains and recorded in the SARAMIS software (Reil *et al.*, 2011). The database was validated with 355 *Clostridium difficile* clinical isolates belonging to 29 different PCR ribotypes. The most frequent ribotypes found during this study were types 001 (70%), 027 (4.8%) and 078/126 (4.7%). The Shimadzu MALDI-TOF MS system with the specially developed SARAMIS database gave the possibility to recognize all three frequent ribotypes and allowed an effective distinction of these strains in due time. As this is the only report about MALDI typing in this area it is still unclear if MS based typing will give comparable results with the PCR ribotyping for *Clostridium difficile*.

It has been shown that *Bacteroides fragilis* strains which harbour the *cfiA* gene (responsible for carbapenemase production in this species) belong to a special subgroup (division II), while those which do not have the *cfiA* gene form division I. The use of various molecular typing methods, such as arbitrary primed PCR, ribotyping, multilocus enzyme electrophoresis and sequencing of the *recA* and *glnA* genes demonstrated that division I and division II of *Bacteroides fragilis* can be clearly distinguished (Gutacker *et al.*, 2000). Two studies using MALDI-TOF MS identification for anaerobic bacteria revealed that the *Bacteroides fragilis* strains belonging to divisions I and II can be differentiated much more rapidly by mass spectrometry than by DNA-based methods. The two studies followed different approaches. Wybo *et al.* (2011) applied the composite correlation index tool of the MALDI Biotyper, and a dendrogram calculated of all tested *Bacteroides fragilis* isolates (248) clearly separated those which harboured the *cfiA* gene. In the other study (Nagy *et al.*, 2011) the MS of well-defined *cfiA*-positive and *cfiA*-negative *Bacteroides fragilis* strains were evaluated by the ClinProTools 2.2 software (Bruker Daltonik). Group-specific peaks and peak shifts were searched for. MS peaks (4826, 9375, 9649 m/z) found to be characteristic for division II isolates were looked for also in a randomly selected group of *Bacteroides fragilis* strains identified earlier by MALDI-TOF MS with high log score (> 2.0). Nine of 38 *Bacteroides fragilis* strains were found to belong to division II and the presence of the *cfiA* gene was confirmed by specific PCR (Nagy *et al.*, 2011). This approach could be used during a six month period in a Hungarian clinical microbiology laboratory when routine identification of *Bacteroides* isolates was done by MALDI-TOF MS. Out of 60 *Bacteroides fragilis* strains, five (8.3%) belonged to division II (harbouring the resistance gene which is responsible for carbapenemase activity in this species). This corresponds to or is slightly higher than the prevalence of these strains (5.7%) observed previously in this country (Fenyvesi *et al.*, 2014). Dendrogram created by MALDI-TOF spectra analysis grouped together the carbapenem-resistant *Bacteroides fragilis* strains in another study too (Trevino *et al.*, 2012). Using this approach it is possible to report carbapenemase positive *Bacteroides fragilis* isolates directly identified from positive blood cultures using the MALDI Biotyper OC software and a dedicated library of *cfiA*-negative and *cfiA*-positive MSPs (Johansson *et al.*, 2014a). Getting the *cfiA*-positive MSP as first best match and with a log score difference of > 0.3 to the second best match was sufficient to consider the isolate *cfiA*-positive. Detection of the *cfiA* gene by PCR confirmed this approach.

Propionibacterium acnes strains have frequently been referred to as a member of the anaerobic skin flora with relatively low pathogenicity. Beside their accepted pathogenic role in acne, *Propionibacterium acnes* isolates are now recognized as pathogens in several different infections such as prosthetic joint infection, post-neurosurgical infection, endocarditis and osteomyelitis. *Propionibacterium acnes* can be differentiated into a number of distinct phylotypes by MLST or other typing methods, known as types IA_1, IA_2, IB, IC, II and III (McDowell *et al.*, 2012). It has been shown that there are associations of the *Propionibacterium aces* phylotypes with certain pathologic conditions. In a recent study (Nagy *et al.*, 2013) the MS-based typing was tested for the resolution of these genetic subgroups of *Propionibacterium acnes* after routine identification by MALDI-TOF MS (Bruker MALDI Biotyper). The ClinProTools 2.2 and the FlexAnalysis 3.3 softwares (Bruker) were used to analyse the mass spectra of reference strains belonging to types IA, IB, IC, II and III. Peak variations between the different types of *Propionibacterium acnes* were identified visually and additionally by FlexAnalysis 3.3 software. A differentiating library was created and used to type clinical isolates of *Propionibacterium acnes*. The MALDI-TOF MS typing results were comparable with those obtained blindly by MLST for the clinical isolates; however, the MS-based typing could not differentiate subtypes in the IA phylogroup of *Propionibacterium acnes* (Nagy *et al.*, 2013). Consequent usage of this typing method directly after identification of *Propionibacterium acnes* isolates by MALDI-TOF MS may help to find further correlation between specific infection processes and the main types of *Propionibacterium acnes*.

Direct identification of anaerobic bacteria from positive blood cultures

Strict anaerobic bacteria are isolated as causative agents of sepsis in about 2–5% of all positive blood cultures (Finegold and George, 1989). Gram stained morphology rarely helps in identification. Subcultures in anaerobic and aerobic environment as well as species determination of the colonies by MALDI-TOF MS are available as rapid methods today. A new approach is the direct identification of bacteria and fungi from positive blood cultures by MALDI-TOF MS (La Scola *et al.*, 2009; Moussaoui *et al.*, 2010; Meex *et al.*, 2012; Leli *et al.*,

2013) using different samples preparation methods (Sepsityper kit – Bruker, or in-house methods). These studies evaluating the direct identification from the positive blood cultures so far have included only very few anaerobic isolates. Using an in-house sample preparation method, Leli *et al.* (2013) reported the species-level identification of 85.5% of Gram-positive and 96.9% of Gram-negative isolates directly from the positive blood cultures, and seven of seven anaerobic species (three *Bacteroides fragilis*, one *Bacteroides thetaiotaomicron*, one *Clostridium paraputrificum*, one *Parvimonas micra* and one *Actinomyces odontolyticus*) were correctly identified if ≥ 1.7 log score was accepted as species level identification. In an other study, 91% of Gram-negative and 89% of Gram-positive bacteria in positive blood cultures were correctly identified at species level including 5 of 13 anaerobic isolates (all *Bacteroides fragilis*) (Moassaoui, *et al.*, 2010). However, one *Bacteroides fragilis*, one *Bacteroides ovatus*, four *Propionibacterium acnes* and two *Clostridium* spp. could not be identified even on genus level. Another study confirmed that *Bacteroides fragilis*, one of the most frequent anaerobic causative agents of sepsis, can be identified directly from the positive blood cultures with high log scores (Johannson, *et al.*, 2014b). Meex *et al.* (2012) evaluated all BacT/ALERT (BioMerieux) anaerobic positive blood cultures using two different sample preparation methods for direct identification. A high percentage of the Gram-negative isolates (including only one *Bacteroides vulgatus* with a negative result) could be identified (82.5% and 90%, respectively); however, only 52% and 58% of Gram-positive bacteria were identified (including one *Actinomyces odontolyticus* with a species level identification by the Sepsityper). Using the Vitek MS system for direct identification of isolates from the positive blood culture bottles a lysis centrifugation method was applied, which involves a four-step washing and centrifugation procedure (Foster, 2013). A total of 253 positive monobacterial blood cultures were tested and 92.1% and 88.1% of the organisms were identified on genus or species level, respectively. Only five bottles contained anaerobic isolates: four *Bacteroides* spp., all identified on species level and one *Leptotrichia* sp., which could not be identified even at genus level.

Further studies are needed to establish the applicability of this method for the rapid identification of a wide range of different anaerobic species obtained directly from positive blood cultures or from the culture of any other, originally sterile body sites. A recent study suggests to subculture the positive blood cultures, incubate them on solid media for five hours and use immediately, when colonies appear on the surface of the media for usual MALDI-TOF MS identification (Verroken *et al.*, 2014). Out of the 10 different anaerobes involved in this study only one *Actinomyces* sp. and a *Clostridium perfringens* were identified by this approach. As anaerobic bacteria are growing rather slowly even in excellent anaerobic environment, the 5 hours incubation time of the subcultures from the positive blood cultures may not be sufficient for anaerobic bacteria.

Use of MALDI-TOF MS for antibiotic resistance determination of anaerobic bacteria

Despite the rapid identification of anaerobic bacteria by mass spectrometry the main criticism from the clinician's side remains that antibiotic susceptibility can not be determined in real time. The only possibility for anaerobes in this moment is to detect the carbapenem hydrolysis by a small subgroup of *Bacteroides fragilis* isolates which harbour the *cfiA* gene and an IS element upstream of the resistance gene, which is needed for the expression of the gene. Johannson *et al.* (2014b) proved that ertapenem as substrate can detect the activity of the carbapenemase gene of *Bacteroides fragilis* within 2.5 h incubation time, if MALDI-TOF measurement compared the ertapenem related peak pattern in culture media with and without the *Bacteroides fragilis* strains. Inhibition of the carbapenemase enzyme with 2,6-pyridinecaroxylic acid was used to prove specificity of the test. Despite of new publications based on experiments with aerobic bacteria (Demirev *et al.*, 2013; Lange *et al.*, 2014) on different approaches to set up antibiotic susceptibility testing by mass spectrometry, at the present time no other proven test is available for anaerobic bacteria.

Use of MALDI-TOF MS to identify other fastidious bacteria

Numerous human pathogenic, aerobic, facultative anaerobic bacteria belong to those which need special nutrition for their culture circumstances and are slow-growing, producing small colonies during 24–48 h of incubation time of the clinical specimens. A special group of these organisms are the members of the HACEK group, species belonging to the genera *Haemophilus, Aggregatibacter, Cardiobacterium, Eikenella* and *Kingella*. They are responsible for several serious infections in paediatric and adult patients including upper and lower respiratory tract infection, endocarditis, meningitis, septic arthritis, osteomyelitis, abdominal abscesses (Goldberg and Katz, 2006). Though the ability to isolate these bacteria has improved greatly by using rich culture media, proper atmosphere for incubation or extending the incubation time of blood culture bottles as well as primary isolation plates, it is still difficult to correctly identify them by conventional biochemical tests. The correct identification of *Haemophilus* isolates on species level may influence selection of antibiotic therapy. *Haemophilus influenzae* causes sever localized respiratory tract infection such as otitis media, sinusitis, bronchitis, pneumonia and often are the causative agents of the acute exacerbation of chronic obstructive pulmonary disease. Non-encapsulated isolates of *Haemophilus influenzae* are difficult to differentiate from the phylogenetically closely related *Haemophilus haemolyticus* isolates, both being members of the normal flora of healthy adults and children. Several studies evaluated the performance of MALDI-TOF MS based identification of these fastidious bacteria at species level and compared the results with those obtained by conventional methods and DNA based molecular techniques (Couturier *et al.*, 2011; Wallet *et al.*, 2011; Boucher *et al.*, 2012; Frickmann *et al.*, 2013; Powell *et al.*, 2013; Zhu *et al.*, 2013; Bruin *et al.*, 2014).

Although the use of MALDI-TOF MS for identification of common human pathogenic bacteria has been already extensively reported, its use for identification of different species of *Haemophilus* was first published quite late (Haag *et al.*, 1998). In this publication out of the known 13 different species the identification of four pathogenic species were evaluated such as *Haemophilus influenzae, Haemophilus parainfluenzae, Haemophilus aphrophilus* and *Haemophilus ducreyi*, responsible for different types of infections and with different pathogenicity, using the PerSeptive Biosystems (Framingham, MA, USA) Voyager-DE MALDI-TOF MS. The 4 species could be clearly separated by the protein peaks in the range of 5000–20,000 m/z. Interestingly, *Haemophilus ducreyi* isolates, obtained from patients suffering from the sexually transmitted genital ulcer (chancroid), could be identified by comparison with the reference strain; however, they could be differentiated in the same time from each other by peaks not found in the reference strain (Haag *et al.*, 1998).

Further studies evaluated the applicability of MALDI-TOF to discriminate two closely related species of *Haemophilus* genus. Using classical methods, the colony morphology and requirements of growth factors, the pathogenic nontypable *Haemophilus influenzae* and the commensal respiratory isolates *Haemophilus haemolyticus* cannot be differentiated. Several time-consuming molecular methods which are not suitable for routine usage, including MLST typing, have been developed in attempts to differentiate the two species. Extensive database developments were carried out for the Bruker system by different laboratories (Zhu *et al.*, 2013; Bruin *et al.*, 2014). In one study the Bruker Biotyper 2.0 database (which did not contain at that time any *Haemophilus haemolyticus* isolates) was challenged by Chinese strains of both species (Zhu *et al.*, 2013). *Haemophilus influenzae* strains gave low log scores and seven of 10 *Haemophilus haemolyticus* isolates were recognized as *Haemophilus influenzae* with log score < 2.0 and three were not identified. After including 10 strictly defined nontypable *Haemophilus influenzae* and 10 *Haemophilus haemolyticus* strains from the collection of the authors the MALDI-TOF MS identified 100% of 42 non-typable *Haemophilus influenzae* isolates on species level, most of them with very high log score (> 2.3) whereas 80% of *Haemophilus haemolyticus* isolates were also identified on species level with the same high log score (Zhu *et al.*, 2013). Altogether 244 *Haemophilus influenzae* and 33 *Haemophilus haemolyticus* isolates obtained from different clinical samples including CFS, blood culture and upper and lower respiratory tract specimens were tested in another study and the MALDI-TOF MS identification results were compared with MLST based identification (Bruin *et al.*, 2014). There was a 99.6% agreement between the two methods to identify these two species showing the rapid discriminatory power of MALDI-TOF MS, if database developments were carried out. Other members of the genus were successfully identified by MALDI-TOF MS such as *Haemophilus pittmaniae* (Boucher *et al.*, 2012) or different isolates of *Haemophilus parainfluenzae* (Frickmann *et al.*, 2013).

Evaluation of the databases and the performance of the two systems (Vitek MS and Bruker MS) (Powell *et al.*, 2013 and Couturier *et al.*, 2011, respectively) for these fastidious, slow growing Gram-negative bacteria were going on, to replace time-consuming, classical identification of species belonging in HACEK group. Vitek MS system correctly identified 100% of 29 *Eikenella corrodens*, 100% of 10 *Aggregatibacter actinomycetemcomitans* and 90% of 20 *Kingella kinge* clinical isolates (Powell *et al.*, 2013). Bruker MS system identified 93% of HACEK organisms correctly to the genus level using the Bruker database, and 100% of the strains were identified to genus level using a custom database that included clinical isolates (Couturier *et al.*, 2011). All these investigations show that despite of the fastidious nature of HACEK organisms MALDI-TOF MS provides an attractive alternative for clinical microbiological laboratories which intend to simplify and accelerate their HACEK identification algorithm.

To summarise the benefits of MALDI-TOF MS in clinical microbiology laboratory practice dealing with fastidious bacteria such as anaerobes and members of the HACEK group

The main advantages of using the MALDI-TOF MS method in the field of anaerobic bacteria are:

1 The mass spectral fingerprint patterns permit us to identify a wide range of human pathogenic anaerobic and other fastidious facultative anaerobic isolates present in mixed

infections. This is eliminating the need of the time-consuming subculturing to prove anaerobic or facultative anaerobic nature of the isolates and achieve enough biomass for classical identification.

2 By the development of databases a great variety of human pathogenic anaerobic bacteria can be identified on species level instead of reporting to the clinicians as 'mixed anaerobic infection'.

3 Different taxa which are non-fermentative, lack of phenotypic markers, and are inactive in biochemical tests, will be accepted and proven as possible pathogens behind some of the infectious processes compared to former time, when routine laboratories often gave up species level identification due to lack of possibilities.

Since the routine application of MALDI-TOF MS in clinical microbiological laboratories has been introduced, case reports prove the existence of rare anaerobic bacteria behind sever infections such as the human pathogenic *Brachispira aalborgi* or *Brachyspira pilosicoli* (Calderaro *et al.*, 2013), *Ruminococcus gnavus* in septic arthritis (Titécat *et al.*, 2014) or the rapid identification of *Clostridium tertium* obtained from the aerobic plate of a subculture of the anaerobic blood culture (Salvador *et al.*, 2013). The timely and correct identification of *Haemophilus* spp. and the slow growing members of the HACEK group have clinical importance to decide the pathogenic role of the isolates (*Haemophilus influenzae* versus *Haemophilus haemolyticus*) and the rapid species level identification of the important pathogenic, fastidious bacteria belonging to the HACEK group.

References

Barba, M.J., Fernández, A., Oviano, M., Fernández, B., Velasco, D., and Bou, G. (2014). Evaluation of MALDI-TOF mass spectrometry for identification of anaerobic bacteria. Anaerobe *30*, 126–128.

Barbuddhe, S.B., Maier, T., Schwarz, G., *et al.* (2008). Rapid identification and typing of *Listeria* species by matrix assisted laser desorption ionization-time of flight mass spectrometry. Appl. Environ. Microbiol. *74*, 5402–5407.

Barreau, M., Pagnier, I., and La Scola, B. (2013). Improving the identification of anaerobes in the clinical microbiology laboratories through MALDI-TOF mass spectrometry. Anaerobe 22, 123–125.

Boucher, M.B., Bedotto, M., Couderc, C., Gomez, C., Reynaud-Gaubert, M., and Drancourt, M. (2012). *Haemophilus pittmaniae* respiratory infection in a patient with siderosis: a case report. J. Med. Case Reports 6, 120.

Bruin, J.P., Kostrzewa, M., van der Ende, A., Badoux, P., Jansen, R., Boers, S.A., and Diederen, B.M.W. (2014). Identification of *Haemophilus influenzae* and *Haemophilus haemolyticus* by matrix-assisted laser desorption ionization-time of flight mass spectrometry. Eur. J. Clin. Microbiol. Infect. Dis. 33, 279–284.

Calderaro, A., Piccolo, G., Montecchini, S., Buttrini, M., Gorrini,C., Rossi, S., Arcangeletti, M.C., De Conto, F., Medici, M.C., and Chezzi, C. (2013). MALDI-TOF MS analysis of human and animal *Brachyspira* species and benefits of database extension. J. Proteomics *78*, 273–280.

Chean, R., Kotsanas, D., Francis, M.J., Palombo, E.A., Jadhav, S.R., Awad, M.M., Lyras, D., Korman, T.M., and Jenkin, G.A. (2014).Comparing the identification of *Clostridium* spp by two Matrix-assisted laser desorption ionization-time of flight (MALDI-TOF) mass spectrometry platforms to 16S rRNA PCR sequencing as a reference standard: A detailed analysis of age of culture and sample preparation. Anaerobe *30*, 85–89.

Claypool, B.M., Yoder, S.C., Citron, D.M., Finegold, S.M., Goldstein, E.J., and Haake, S.K. (2010). Mobilization and prevalence of a fusobacterial plasmid. Plasmid *63*, 11–19.

Coltella, L., Mancinelli, L., Onori, M., *et al.* (20013). Advancement in the routine identification of anaerobic bacteria by MALDI-TOF mass spectrometry. Eur. J. Clin. Microbiol. Infect. Dis. *32*, 1138–1192.

Copsey, S., Scotford, S., Jones, S., Wootton, M., Hall, V., and Howe, R.A. (2013). Evaluation of MALDI-TOF MS for the identification of anaerobes. 23rd ECCMID Berlin, Germany P-1715.

Couturier, M.R., Mehinovic, E., Croft, A.C., and Fisher, M.A. (2011). Identification of HACEK clinical isolates by matrix-assisted laser desorption ionization-time of flight mass spectrometry. J. Clin. Microbiol. *49*, 1104–1106.

Culebras, E., Rodriguez-Avial, I., Betriu, C., Gomez, M., and Picazo, J.J. (2012). Rapid identification of clinical isolates of *Bacteroides* species by matrix assisted laser-desorption/ionization time-of-flight mass spectrometry. Anaerobe *18*, 163–165.

Demirev, P.A., Hagan, N.S., Antoine, M.D., Lin, J.S., and Feldman, A.B. (2013). Establishing drug resistance in microorganisms by mass spectrometry. J. Am. Soc. Mass Spectrom. *24*, 1194–1201.

Federko, D.P., Drake, S.L., Stock, F., and Murray, P.R. (2012). Identification of clinical isolates of anaerobic bacteria using matrix-assisted laser desorption ionization-time of flight mass spectrometry. Eur. J. Clin. Microbiol. Infect. Dis. *31*, 2257–2262.

Fenyvesi, V.S., Urbán, E., Bartha, N., Ábrok, M., Kostrzewa, M., Nagy, E., Minárovics, J., and Sóki, J. (2014). Use of MALDI-TOF/MS for routine detection of *cfiA*-positive *Bacteroides fragilis* strains. Intern. J. Antimicrob. Agents *44*, 469–476.

Finegold, S.M., and George, W.L. (1989). Anaerobic Infections in Humans (Academic Press, San Diego, CA).

Foster, A.G.W. (2013). Rapid identification of microbes in positive blood cultures by use of the Vitek MS matrix assisted laser desorption ionization-time of flight mass spectrometry system. J. Clin Microbiol. *51*, 3717–3719.

Fournier, R., Wallet, F., Grandbastien, B., Dubreuil, L., Courcol, R., Neut, C., and Dessein, R. (2012). Chemical extraction versus direct smear for MALDI-TOF MS spectrometry for identification of anaerobic bacteria. Anaerobe *18*, 294–297.

Frickmann, H., Christner, M., Donat, M., Berger, A., Essig, A., Pobielski, A., Hagen, R.M., and Popper, S. (2013). Rapid discrimination of *Haemophilus influenzae, H. parainfluenzae* and *H. haemolyticus* by fluorescence in situ hybridization (FISH) and two matrix–assisted laser-desorption-ionization time-of-flight mass spectrometry (MALDI-TOF-MS) platforms. PloS One *8*, e63222.

Garner, O., Mochon, A., Branda, J., Burnham, C.A., Bythrow, M., Ferraro, M., Ginocchio, C., Jennemann, R., Manji, R., Procop, G.W., *et al.* (2014). Multi-centre evaluation of mass spectrometric identification of anaerobic bacteria using VITEK MS system. Clin. Microbiol. Infect. *20*, 335–339.

Goldberg, M.H., and Katz, J. (2006). Infective endocarditis caused by fastidious oropharyngeal HACEK micro-organisms. J. Oral Maxillofac. Surg. *64*, 969–971.

Grosse-Herrenthey, A., Maier, T., Gessler, F., Schaumann, R., Böhnel, H., Kostrzewa, M., and Krüger, M. (2008). Challenging the problem of clostridial identification with matrix-assisted laser desorption and ionization time-of-flight mass spectrometry. Anaerobe *14*, 242–249.

Gutacker, M., Valsangiacomo, C., and Piffaretti, J.C. (2000). Identification of two genetic groups in *Bacteroides fragilis* by multilocus enzyme electrophoresis: distribution of antibiotic resistance (*cfiA*, *cepA*) and enterotoxin (*bft*) encoding genes. Microbiol. *146*, 1241–1254.

Haag, A.M., Taylor, S.N., Johnston, K.H., and Cole, R.B. (1998). Rapid identification of speciation of *Haemophilus* bacteria by matrix-assisted laser desorption/ionization time-of-flight mass spectrometry. J. Mass Sectrom. *33*, 750–756.

Hettick, J.M., Kashon, M.L., Slaven, J.E., Ma, Y., Simpson, J.P., Siegel, P.D., Mazurek, G.N., and Weissman, D.N. (2006). Discrimination of intact mycobacteria at the strain level: A combined MALDI-TOF MS and biostatistic analysis. Proteomics *6*, 6416–6425.

Hsu, Y.S., and Burnham, C.D. (2014). MALDI-TOF MS identification of anaerobic bacteria: assessment of pre-analytical variables and specimen preparation techniques. Diagn. Microbiol. Infect. Dis. *79*, 144–148.

Jamal, W.Y., Shahin, M., and Rotimi, V.O. (2013). Comparison of two matrix-assisted laser desorption ionization-time of flight (MALDI-TOF) mass spectrometry methods and API20AN for identification of clinically relevant anaerobic bacteria. J. Med. Microbiol. *62*, 540–544.

Johannson, A., Nagy, E., and Sóki, J; on behalf of ESGAI (2014a). Detection of carbapenemase activities of *Bacteroides fragilis* strains with matrix-assisted laser desorption ionization-time of flight mass spectrometry (MALDI-TOF MS). Anaerobe *26*, 49–52.

Johannson, A., Nagy, E., and Sóki, J. (2014b). Instant screening and verification of carbapenemase activity in *Bacteroides fragilis* in positive blood culture, using matrix-assisted laser desorption ionization-time of flight mass spectrometry. J. Med. Microbiol. *63*, 1105–1110.

Jousimies-Somer, H.R., Summanen, P., Citron, D.M., Baron, E.J., Wexler, H.M., and Finegold, S.M. (2002). Wadsworth-KTL Anaerobic Bacteriology Manual, 6th edn (Star Publishing Company, Redwood City, CA).

Justesen, U.S., Holm, A., Knudsen, E., Andersen, L.B., Jensen, T.G., Kemp, M., Skov, M.N., Gahrn-Hansen, B., and Møller, J.K. (2011). Species identification of clinical isolates of anaerobic bacteria: a comparison of two matrix-assisted laser desorption ionization – time of flight mass spectrometry systems. J. Clin. Microbiol. *49*, 4314–4318.

Kierzkowska, M., Majewska, A., Kuthan, R.T., Sawicka-Grzelak, A., and Mlynarczyk, G. (2013). A comparison of Api20A vs MALDI-TOF MS for routine identification of clinically significant anaerobic bacterial strains to the species level. J. Microbiol. Methods *92*, 209–212.

Knoester, M., van Veen, S.Q., Vlaas, E.C.J., and Kuijper, E.J. (2012). Routine identification of clinical isolates of anaerobic bacteria: matrix-assisted laser desorption ionization-time of flight mass spectrometry performs better than conventional identification methods. J. Clin. Microbiol. *50*, 1504.

Lange, C., Schubert, S., Jung, J., Kostrzewa, M., and Sperbier, K. (2014). Quantitative matrix-assisted laser desorption ionization-time of flight mass spectrometry for rapid resistance detection. J. Clin. Microbiol. *52*, 4155–4162.

La Scola, B., and Raoult, D. (2009). Direct identification of bacteria in positive blood culture bottles by matrix assisted laser desorption ionisation time-of-flight mass spectrometry. PLoS One *4*, E8041.

La Scola, B., Fournier, P.-E., and Raoult, D. (2011). Burden of emerging anaerobes in the MALDI-TOF and the 16S rRNA gene sequencing era. Anaerobe *17*, 106–112.

Latigue, M.F., Hery-Arnaud, G., Haguenoer, E., Domelier, A.S., Schmit, P.O., van der Mee-Marquet, N., Lanotte, P., Mereghetti, L., Kostrzewa, M., and Quentin, R. (2009). Identification of *Streptococcus agalactiae* isolates from various phylogenetic lineages by matrix-assisted laser desorption ionization-time of flight mass spectrometry. J. Clin. Microbiol. *47*, 2284–2287.

Lee, W., Kim, M., Yong, D., Jeong, S.H., Lee, K., and Chog, Y. (2015). Evaluation of VITEK mass spectrometry (MS), a matrix-assisted laser desorption ionization time-of-flight MS system for identification of anaerobic bacteria. Ann. Lab. Med. *35*, 69–75.

Leli, C., Cenci, E., Cardaccia, A., Moretti, A., D'Alò, F., Pagliochini, R., Barcaccia, M., Farinelli, S., Vento, S., Bistoni, F., *et al.* (2013). Rapid identification of bacterial and fungal pathogens from positive blood cultures by MALDI-TOF MS. Int. J. Med. Microbiol. *303*, 205–209.

Martiny, D., Busson, L., Wybo, I., Haj, R.A.E., Dediste, A., and Vandenberg, O. (2012). Comparison of the Microflex LT and Vitek MS systems for the routine identification of bacteria by matrix-assisted laser desorption-ionization time-of-flight mass spectrometry. J. Clin. Microbiol. *50*, 1313–1325.

McDowell, A., Barnard, E., Nagy, I., Gao, A., Tomida, S., Li, H., Eady, A., Cove, J., Nord, C.E., and Patrick, S. (2012). An expanded multilocus sequence typing scheme for *Propionibacterium acnes*: Investigation of 'pathogenic', 'commensal' and antibiotic resistant strains. PLoS One 7, e41480.

Meex, C., Neuville, F., Descy, J., Huynen, P., Hayette. M.-P., De Mal, P., and Melin, P. (2012). Direct identification of bacteria from BacT/ALERT anaerobic positive blood cultures by MALDI-TOF MS: MALDI Sepsityper kit versus an in-house saponin method for bacterial extraction. J. Med. Microbiol. *61*, 1511–1516.

Moussaoui, W., Jaulhac, B., Hoffmann, A.M., Ludes, B., Kostrzewa, M., Riegel, P., Prévost, G., *et al.* (2010). Matrix assisted laser desorption ionization time-of-flight mass spectrometry identifies 90% of bacteria from blood culture vials. Clin. Microbiol. Infect. *16*, 1631–1638.

Nagy, E., Maier, T., Urban, E., Terhes, G., and Kostrzewa, M. (2009). Species identification of clinical isolates of *Bacteroides* by matrix-assisted laser-desorption/ionization time-of-flight mass spectrometry. Clin. Microbiol. Infect. *15*, 796–802.

Nagy, E., Becker, S., Sóki, J., Urbán, E., and Kostrzewa, M. (2011). Differentiation of division I (*cfiA*-negative) and division II (*cfiA*-positive) *Bacteroides fragilis* strains by matrix assisted laser desorption ionization-time of flight mass spectrometry. J. Med. Microbiol. *60*, 1584–1590.

Nagy, E., Becker, S., Kostrzewa, M., Barta, N., and Urbán, E. (2012). The value of MALDI-TOF MS for the identification of clinically relevant anaerobic bacteria in routine laboratories. J. Med. Microbiol. *61*, 1393–1400.

Nagy, E., Urban, E., Becker, S., Kostrzewa, M., Vörös, A., Hunyadkürti, J., and Nagy, I. (2013). MALDI-TOF MS fingerprinting facilitates rapid discrimination of phylotypes I, II and III of *Propionibacterium acnes*. Anaerobe *20*, 20–26.

Nagy, E. (2014). Matrix-assisted laser desorption/ionization time-of-flight mass spectrometry: a new possibility for the identification and typing of anaerobic bacteria. Future Microbiol. *9*, 217–233.

Powell, A.E., Blecker-Shelly, D., Montgomery, S., and Mortensen, J.E. (2013). Application of matrix-assisted laser desorption ionization-time of flight mass spectrometry for identification of the fastidious pediatric pathogens *Aggregatibacter, Eikenella, Haemophilus,* and *Kingella*. J. Clin. Microbiol. *51*, 3862–3864.

Reil, M., Erhard, M., Kuijper, E.J., *et al.* (2011). Recognition of *C. difficile* PCR-ribotypes 001, 027, and 126/078 using an extended MALDI-TOF MS system. Eur. J. Clin. Microbiol. Infect. Dis. *30*, 1431–1436.

Salvador, F., Porte, L., Durán, L., Marcotti, A., Pérez, J., Thompson, L., Noriega, L.M., Lois, V., and Weitzel, T. (2013). Breakthrough bacteremia due to *Clostridium tertium* in a patient with neutropenic fever, and identification by MALDI-TOF mass spectrometry. Int. J. Infect. Dis. *17*, e1062-e1063.

Schmitt, B.H., Cunningham, S.A., Dailey, A.L., Gustafson, D.R., and Patel, R. (2013). Identification of anaerobic bacteria by Bruker Biotyper matrix-assisted laser desorption ionization-time of flight mass spectrometry with on-plate formic acid preparation. J. Clin. Microbiol. *51*, 782–786.

Shah, H.N., Keys. C.J., Schmid, O., and Gharbia, S.E. (2002). Matrix-assisted laser desorption/ionization time-of-flight mass spectrometry and proteomics: a new era in anaerobic microbiology. Clin. Infect. Dis. *35*, 58–64.

Stingu, C.S., Rodloff, A.C., Jentsch, H., Schaumann, R., and Eschrich, K. (2008). Rapid identification of oral anaerobic bacteria cultivated from subgingival biofilm by MALDI-TOF MS. Oral Microbiol. Immun. *23*, 372–376.

Stingu, C.S., Borgmann, T., Rodloff, A.C., Vielkind, P., Jentsch, H., Schellenberger, W., and Eschrich, K. (2015). Rapid identification of oral *Actinomyces* species cultivated from subgingival biofilm by MALDI-TOF-MS. J. Oral Microbiol. *16*, 26110.

Tanigawa, K., Kawabata, H., and Watanabe, K. (2010). Identification and typing of *Lactococcus lactis* by matrix assisted laser desorption ionization-time of flight mass spectrometry. Appl. Environ. Microbiol. *76*, 4055–4062.

Titécat, M., Wallet, F., Vieillard, M.-H., Courcol, R.J., and Loiez, C. (2014). *Ruminococcus gnavus*: An unusual pathogen in septic arthritis. Anaerobe *30*, 159–160.

Trevino, M., Areses, P., Penalver, M.D., Cortizo, S., Pardo, F., del Molino, M.L., García-Riestra, C., Hernández, M., Llovo, J., and Regueiro, B.J. (2012). Susceptibility trends of *Bacteroides fragilis* group and characterization of carbapenemase producing strains by automated REP-PCR and MALDI-TOF. Anaerobe *18*, 37–43.

Veloo, A.C.M., Erhard, M., Welker, M., Welling, G.W., and Degener, J.E. (2011a). Identification of gram-positive anaerobic cocci by MALDI-TOF Mass Spectrometry. Syst. Appl. Microbiol. *34*, 58–62.

Veloo, A.C.M., Knoester, M., Degened, J.E., and Kuijper, E.J. (2011b). Comparison of two matrix-assisted laser desorption ionization-time of flight mass spectrometry methods for the identification of clinically relevant anaerobic bacteria. Clin. Microbiol. Infect. *17*, 1501–1506.

Veloo, A.C.M., Elgersma, P.E., Fridrich, A.W., Nagy, E., and von Winkelhoff, A.J. (2014). The influence of incubation time, sample preparation and exposure to oxygen on the quality of the MALDI-TOF MS spectrum of anaerobic bacteria. Clin. Microbiol. Infect. *20*, O1091–O1097.

Verroken, A., Defourny, L., Lechgar, L., Magnette, A., Delmée, M., and Glupczynski, Y. (2014). Reducing time to identification of positive blood cultures with MALDI-TOF MS analysis after a 5-h subculture. Eur. J. Clin. Microbiol. Infect. Dis. *34*, 405–413.

Wallet, F., Loiez, C., Decoene, C., and Courcol, R. (2011). Rapid identification of *Cardiobacterium hominis* by MALDI-TOF Mass spectrometry during infective endocarditis. Jpn. J. Infect. *64*, 327–329.

Watanabe, Y., Nagai, F., Morotomi, M., Sakon, H., and Tanaka, R. (2010). *Bacteroides clarus* sp. nov., *Bacteroides fluxus* sp. nov. and *Bacteroides oleiciplenus* sp. nov., isolated from human faeces. Int. J. Syst. Evol. Microbiol. *60*, 1864–1869.

Wolters, M., Rohde, H., Maier, T., Belmar-Campos, C., Franke, G., Scherpe, S., Aepfelbacher, M., and Christner, M. (2011). MALDI-TOF MS fingerprinting allows for discrimination of major methicillin resistant *Staphylococcus aureus* lineages. Inter. J. Med. Microbiol. *301*, 64–68.

Wybo, I., De Bel, A., Soetens, O., Echahidi, F., Vandoorslaer, K., Van Cauwenbergh, M., and Piérard, D. (2011). Differentiation of *cfiA*-negative and *cfiA*-positive *Bacteroides fragilis* isolates by matrix assisted laser desorption ionization – time of flight mass spectrometry. J. Clin. Microbiol. *49*, 1961–1964.

Wybo, I., Soetens, O., De Bel, A., Echahidi, F., Vancutsem, E., Vandoorslaer, K., and Piérard, D. (2012). Species identification of clinical *Prevotella* isolates by matrix assisted laser desorption ionization time-of-flight mass spectrometry. J. Clin. Microbiol. *50*, 1415–1418.

Zhu, B., Xiao, D., Zhang, H., Zhamg, Y., Gao, Y., Xu, L., Lv, J., Wang, Y., Zhang, J., and Shao, Z. (2013). MADI-TOF MS direct differentiates nontypable *Haemophilus influenzae* from *Haemophilus haemolyticus*. PLoS One 8, e56139.

Identification, Typing and Susceptibility Testing of Fungi (Including Yeasts) by MALDI-TOF MS

3

Anna Kolecka, Maurizio Sanguinetti, Teun Boekhout and Brunella Posteraro

Abstract

Matrix-assisted laser desorption ionization-time of flight mass spectrometry (MALDI-TOF MS) has developed into a robust tool for the routine identification of many microbes, including bacteria, yeasts and filamentous fungi, which remain relevant for maintaining the quality of the human lifestyle, e.g. related to health, food, industry, ecology, water quality, waste management, and others. The number of correct and erroneous results depends mainly on the sample preparation method, the number of reference spectra present in the databases and the cut-off values used. These studies confirmed that MALDI-TOF MS platform outperformed traditional identification methods for fungi. Examples will be given on the use of MALDI-TOF MS for the identification of yeasts and filamentous fungi, including strain typing and susceptibility testing.

Introduction

Rapid identification of isolates of yeasts and filamentous fungi is important in many areas of laboratory diagnostics, such as clinical microbiology, industrial microbiology, food safety, pest control, quarantine services, ecology, and many others. Until recently identification of yeast isolates has been mainly achieved by culture-based and DNA-based methods. The microbiological and physiological assays were and are widely used in the routine diagnostic practice. The use of auxanograms, test tubes with chemically defined growth media and fermentations tests (Kurtzman *et al.*, 2011) were widely applied in the identification of yeasts. However, also precast panels of assimilation/growth tests using defined sets of carbon and nitrogen compounds, such as commercially available kits (e.g. API 20C AUX, ID 32C and RapID Yeast Plus system) are still widely applied and can be automated to a large extent. Lately, the automated Vitek Systems 1 and 2 (bioMérieux, France) have increasingly been used for the routine identification of yeasts and filamentous fungi. The identification of filamentous fungi relied mainly on morphological expertise. More recently, sequence analysis of parts of the ribosomal DNA (rDNA), such as the D1/D2 domains of the large subunit (LSU, 26S or 28S) rDNA and the internally transcribed spacers 1 and 2 (ITS1/ITS2), so-called molecular barcodes, largely replaced the phenotypic methods for fungal identification (White *et al.*, 1985; Kurtzman and Robnett, 1998; Fell *et al.*, 2000; Schoch et al., 2012). The utilization of molecular barcodes has to a large extent improved the reliability of fungal identification. Unfortunately, some of the most commonly used barcodes, ITS and D1/D2, suffer from a lack of resolution in many fungal genera, such as *Aspergillus*, *Penicillium*, *Trichoderma*, *Cladosporium* and *Alternaria*, just to name a few. In those groups additional genes or parts thereof need to be sequenced to obtain a reliable species identification. As the genes of choice differ from genus to genus, some pre-identification of the isolates is needed, which again requires some morphological knowledge of fungi. Another limitation is the quality of sequence data stored in big repositories, such as GenBank, that is not always reliable (http://www.ncbi.nlm.nih.gov/genbank) (Vilgalys, 2003; Nilsson *et al.*, 2006), and also missing data in the databases (Romanelli *et al.*, 2010).

Application of matrix-assisted laser desorption ionization-time of flight mass spectrometry (MALDI-TOF MS) offers new perspectives in the identification of yeasts and filamentous fungi (Marklein *et al.*, 2009; Cassagne *et al.* 2011; Iriart *et al.*, 2012; Seyfarth *et al.*, 2012; Posteraro *et al.*, 2013; Sendid *et al.*, 2013). Since the 1980s, the MALDI-TOF MS technology underwent technical improvements (Karas *et al.*, 1987, Tanaka *et al.*, 1988) and the first commercial instruments were introduced in the early 1990s. These instruments were mainly used for the identification and differentiation of bacterial species (Hillenkamp and Karas, 1990; Claydon *et al.*, 1996; Holland *et al.*, 1996; Krishnamurthy *et al.*, 1996). After these initial successes, MALDI-TOF MS systems have revolutionized microbial identification in many laboratories worldwide. In this review we discuss the prospects of MALDI-TOF MS-based identification of yeasts and filamentous fungi, the methods in use, assessment of antifungal susceptibility properties, and we will sketch some future developments.

MALDI-TOF MS systems in use

Currently, the MALDI-TOF MS approach is commercialized by a number of manufacturers, namely MALDI Biotyper (Bruker Daltonics, Germany), Vitek MS (bioMérieux, France), Axima (Shimadzu)-SARAMIS and Andromas (Andromas, France) (Bille *et al.*, 2012, Bader, 2013; Lacroix *et al.*, 2013; Lohmann *et al.*, 2013). The performance of two MALDI-TOF MS systems, namely MALDI Biotyper (Bruker Daltonics, Germany) and Andromas (Andromas, France), were compared using procedures recommended by the manufacturers (Bader *et al.*, 2011; Lacroix *et al.*, 2013). Additionally, two laboratories performed independently a comparative evaluation of the Vitek MS (bioMérieux, France) and MALDI Biotyper (Bruker Daltonics, Germany) systems for the identification of clinically significant yeasts (Mancini *et al.*, 2013; Jamal *et al.*, 2014). Both studies reported similar identification rates between both systems when using commercially available databases. Better performance of the MALDI Biotyper system was noted with an in-house extended database (Mancini *et al.*, 2013). Although these studies confirmed equal performance of both platforms for yeasts identification and outperformed traditional identification methods, the sensitivity and specificity of identification results of yeasts turned out to be variable and difficult to compare between laboratories as they used different sample preparation methods and most studies were performed and presented by a single laboratory.

The first multicentre evaluation of Vitek MS (bioMérieux, France) that enrolled five diagnostic clinical microbiology laboratories from North America indicated that the Vitek MS (bioMérieux, France) system yielded identification rates comparable or superior to both traditional biochemical and sequence-based identification systems, stressing out that some strains were not identified ($n = 24$, 2.8%) or misidentified ($n = 5$, 0.6%) (Westblade *et al.*, 2013). Recently, an inter-laboratory study was established in which the performance of the MALDI Biotyper (Bruker) regarding the identification of yeasts was investigated in a clinical setting (Vlek *et al.*, 2014). This multicentre study compared three methods of sample preparation, use of various cut-off values, use of two databases, and evaluated the relationship between the number of entries in the database and the number of correct identifications. In conclusion, it was suggested that MALDI-TOF MS is a reliable system for identifying species of commonly, but also rarely occurring pathogenic yeasts, and that the threshold value can be lowered to 1.7. Significant impact on the accuracy of identifications remained the choice of sample preparation method and the number of entries in the database. Further investigation is needed to optimize the direct transfer method for the routine identification of yeasts and filamentous fungi.

To date, MALDI-TOF MS allows reliable identification of fungi in a short turnaround time. Although the purchasing of a MALDI-TOF MS instrument remains relatively costly, the routine use of chemicals and other reagents is rather inexpensive. Proper choice of the method used for sample preparation and further optimization thereof may decrease costs and hands-on time per sample, thus reducing turnaround time.

MALDI-TOF MS identification of yeast isolates

MALDI-TOF MS-based identification of yeasts has been applied mainly in clinical laboratories, but also in food science and ecology (e.g. Marklein *et al.*, 2009; Dhiman *et al.*, 2011; Seyfarth *et al.*, 2012; Krause *et al.*, 2013; Sendid *et al.*, 2013; Usbeck *et al.* 2013; Vallejo *et al.*, 2013; Agustini *et al.*, 2014; Pavlovic *et al.*, 2014). In comparison with other identification methods, such as sequence analysis of the D1/D2 domains of the LSU and the ITS1/ITS2 regions of the rDNA, MALDI-TOF MS is able to provide accurate identifications of yeasts in short turnaround times (Marklein *et al.*, 2009; Dhiman *et al.*, 2011; Tan *et al.*, 2013). As with all microbial identification methods, the reliability of a method not only depends on the technical reproducibility and sensitivity, but also on the extent of the databases (Stevenson *et al.*, 2010; Mancini *et al.*, 2013; Kolecka *et al.*, 2014).

Sample preparation methods

In routine practice, the quality of identifications achieved depends mainly on the sample preparation method, the number of species and reference spectra in a database, the cut-off values used and the human factor (technician expertise and proficiency, hands-on experience time, and other technicalities such as correct alignment of the machine, etc.) (Vlek *et al.*, 2014).

The sample preparation method has a major impact on the accuracy of identifications. Three methods are in use for MALDI Biotyper identifications: (i) direct transfer (DT), (ii) extended direct transfer (eDT), and (iii) ethanol (EtOH)-formic acid (FA) extraction (MALDI Biotyper user manual, Bruker Daltonics). In the DT method, a thin smear of biological material is placed onto a target plate, which is overlaid with 1 µl of α-cyano-4-hydroxycinnamic acid (HCCA) matrix solution prepared according to the protocol of the manufacturer. In the eDT method, the biomass is treated with 1 µl of 70% FA on the target plate prior to the HCCA overlay. In the EtOH-FA method, one or two loops of yeast biomass (1 µl volume, sterile inoculation loop) are used for crude protein extraction as described previously (Kolecka *et al.*, 2013). One microlitre of the crude protein extract is spotted onto the target plate, and after air-drying, overlaid with 1 µl of HCCA matrix solution. For all methods tested strains are usually spotted in duplicate but recently, procedure with one spot only becomes more common. The target plate provided by Bruker is reusable after harsh chemical cleaning using 80% aqueous trifluoroacetic acid (TFA) (MALDI Biotyper user manual, Bruker Daltonics).

The Vitek MS (bioMérieux, France) system has two sample preparation methods, namely direct deposit of yeasts (DT) and an extraction method (EtOH-FA) for moulds. In the DT method biomass is directly applied on two spots of the disposable target plate and lysed with 0.5 µl of 25% FA and when dried overlaid with ready-to-use HCCA. In the EtOH-FA, a fragment of a mould colony is collected with a wet swab and suspended in 300 µl sterile deionized water followed by 900 µl absolute ethanol. After centrifugation, the pellet is resuspended in 40 µl of 70% FA and 40 µl of acetonitrile (ACN). After centrifugation, 1 µl of the supernatant is deposited on the target plate in duplicate and when dried overlaid with 1 µl of HCCA. For both MALDI-TOF MS platforms (Bruker and bioMérieux), these methods do not vary much and may well be applied for bacteria and yeasts, and with some modifications also for filamentous fungi.

Significant differences between labs using all three methods were reported recently when identifying yeasts (Vlek *et al.*, 2014), suggesting that at present the choice of the sample preparation method has a significant impact on the accuracy and reliability of identifications in routine laboratory practice. Other studies compared the various preparation methods but these studies were done by a single laboratory only (Bader *et al.*, 2011; Cassagne *et al.*, 2013). Sensitivity and specificity of identification results of yeasts reported worldwide are variable and difficult to compare between laboratories as they report different identification rates by using different sample preparation methods, ranging from 16% (Pinto *et al.*, 2011) to approximately 21% (Cassagne *et al.*, 2013) when using the direct transfer method, to more than 90% of correct identifications with the formic acid/ethanol extraction method (Marklein *et al.*, 2009; Bader *et al.*, 2011; McTaggart *et al.*, 2011; Vlek *et al.*, 2014). Importantly, improvement of reproducible transfer of biomass or inoculum spotting onto a MALDI-TOF MS target plate is needed. This step remains labour intensive, because it is still performed manually and it results in low precision and significant variation in target plate preparation. This may be a challenging task bearing in mind differences between growth of various yeasts and filamentous fungi, e.g. size of the colony, that may impact the amount of biomass, the colony structure (smooth, wrinkled, warty, woolly, slimy, growing into agar, dragging, etc.) or presence of inhibitors, such as melanin produced by black yeasts (Buskirk *et al.*, 2011).

In the routine laboratory practice the use of the DT or eDT methods are favoured for identification as the full extraction procedure is rather laborious and time-consuming (Gorton *et al.*, 2014). Most of the labs are currently using polished steel target plate (MSP 96 target polished steel; Bruker, Bremen, Germany) independent on the sample preparation method they have chosen to follow. Alternatively, the ground steel target plate (MSP 96 target ground steel; Bruker, Bremen, Germany) is recommended by the manufacturer to be used for the DT and/or eDT while polished steel target plate should be applied for the samples prepared with EtOH/FA extraction method. In a recent study, five hospitals compared the utility of those two target plates for identification of eight *Candida* species using DT (Riat *et al.*, 2015). It was confirmed that for the identification of 206 clinical yeasts together with a bias testing set of 15 isolates, the same outcome of identification results can be expected when using the DT method using ground steel plates as for when using eDT on polished steel target plates, thus suggesting that ground steel plates should be enrolled for the yeast identification when using the DT method.

Growth conditions (recommendations)

The yeast strains to be used for MALDI-TOF MS identifications are usually cultured on Sabouraud glucose agar (SGA) plates, with some modifications depending on the growth requirements of the fungus or standard procedures in specific laboratories. Use of different media, namely Sabouraud glucose agar (SGA), that allows capsule formation, and SGA + 0.5 NaCl that inhibits capsule formation was shown to have no impact on MALDI-TOF MS identifications of *Cryptococcus neoformans* (Hagen *et al.*, 2015). For lipophilic *Malassezia* yeasts, the use of non-lipid supplemented media, like SGA, will not work because fatty acid supplementation is needed for these lipophilic or lipid dependent yeasts. For that reason, specialized media should be used [e.g. modified Dixon's agar (mDA) or modified Leeming and Notman agar (MLNA)]. The use of mDA plates was recently recommended by Kolecka *et al.* (2014) because oil residues may be transferred from the surface of MLNA plates together with the yeast biomass, which makes it difficult to dry and measure on the spots on the target plate. In order to prevent some of the strains from growing into the agar sterile polycarbonate PCTE filtration membranes (76 mm in diameter size, 0.1 mm pore size, GE Water and Process Technologies) can be applied on the agar surface, on which the fungal material is inoculated (Kolecka *et al.*, 2013). Such modification was confirmed to work well for some of the dimorphic yeast strains, (e.g. *Saprochaete* spp. or *Trichosporon* spp.), especially for those species that exhibit cerebriform and a radial furrows at the surface of the colonies, or those that become membranous with age. Melanin production in black fungi, such as *Aspergillus niger*, was suppressed by adding tricyclazole to the fungal growth medium (Buskirk *et al.*, 2011). Also, standardized growth conditions and media for food-borne yeasts are necessary as different culturing conditions may affect the acquired mass spectra fingerprints. Many media are in use to isolate yeasts from food products, but media such as malt extract agar (MEA) or dichloran rose Bengal agar (DRBC) are recommended (Samson *et al.*, 2010). Usbeck *et al.* (2013) tested cultures that grew on five different media, namely yeast extract–peptone–glucose (YPG), yeast–malt–peptone (YM), malt extract (ME), Sabouraud glucose and wort (malt–maltose–dextrin–casein peptone), but cultures from YPG and YM showed the best performance of MALDI-TOF MS with the lowest standard deviation for identification hit rates. It was also advised to work with fresh, young cultures of yeasts (approx. 20 hours) (Usbeck *et al.*, 2013). Yeast cultures older than > 24 hours' incubation often resulted in problems with obtaining IDs and this has a negative effect on the application of the DT and eDT methods (Kolecka *et al.*, 2013; Vlek *et al.*, 2014).

Species-specific score cut-off values, repeats

In the identification process the generated data can be categorized into (i) correct identification at the genus or species level, (ii) not reliable identification (NRI), (iii) misidentification at the genus (major error), (iv) no peaks found (NPF) or (v) no growth (NG). Independent on the sample preparation methods or database used, MALDI-TOF MS identification results are automatically classified using the log-score values according to the manufacturer's recommended criteria: (1) correct genus and species identification (score value ≥ 2.0), (2) secure genus identification (score value 1.7–2.0) and (3) no reliable identification (score value < 1.7). Generally, the identification is considered correct if at least one spot from a duplicate gives a reliable identification with a score value > 2.0. Recently, many studies evaluated the correctness of the yeast identifications with score values > 1.7. Based on a multicentre study, a cut-off value of 1.7 was considered appropriate for yeast identification in clinical laboratories, as this study showed that the majority of misidentification results were associated with scores below 1.7 (NRI), regardless of the sample preparation method and database used (Vlek *et al.*, 2014). This is in agreement with previous studies that reported an increase in the number of accurate identifications at species level when lowering the score threshold to the value of 1.7 or 1.8 (Stevenson *et al.*, 2010; Dhiman *et al.*, 2011; van Herendael *et al.*, 2012; Rosenvinge *et al.*, 2012). The MALDI-TOF MS Biotyper system did not provide false positive identifications with scores > 1.7 for fungal species or genera (Vlek *et al.*, 2014). It was shown that MALDI-TOF MS identifications generated with scores > 1.7 were conclusive and concordant with results obtained by molecular barcodes and thus it was proposed that the species-specific score cut-off values of clinically relevant yeasts can be lowered to 1.7. As a consequence, these latest data made the manufacturer to decide to implement modifications in the interpretation of score values, namely log score values (≥ 2.0) are recognized as 'high confidence identifications' and scores (1.7–2.0) as 'low confidence identification' (MALDI Biotyper user manual, Bruker Daltonics). The samples of 'no identification' usually fall into the category of 'not reliable identifications' that were achieved with score values (< 1.7) or 'no peaks found' as a consequence of a flat line of the mass spectrum.

Databases

Presently (March, 2015), the standard MALDI Biotyper microorganism database (BDAL, Bruker Daltonics) contains 5627 main spectra (MSPs) of 380 microorganism genera covering more than 2200 species. This database comprises 649 MSPs for fungi, of which 183 MSPs cover *Basidomycota* with 156 MSPs of *Agaricomycotina*, and 464 MSPs of *Ascomycota* with 399 MSPs of *Saccharomycotina*. The yeast panel in this BDAL database contains 574 MSPs representing 172 species of 28 genera. An additional database of filamentous fungi (Fungi Library 1.0 – #700281, Bruker Daltonics) contains 365 MSPs of more than 110 species from approximately 40 different genera. The current Vitek MS (IVD) spectral database MS-ID version 2 from the other commercial diagnostic platform (bioMérieux, France) contains 29,873 reference spectra of 751 microbial species, including 1266 spectra of different fungi, covering 110 species of different yeasts and moulds (personal communication, Valérie Monnin and Alex van Belkum, bioMérieux, France). In the following version, the number of spectra in the database will increase to 38428 and the fungal panel will expand with 164 new species of yeasts and moulds. Over the last decade, the reliability of identification of commonly and rarely found yeasts improved considerably by adding new MSPs to the database, because a higher number of reference MSPs per species increased the biological diversity in the database. The usefulness of database expansion in order to reliably identify emerging rare pathogenic yeast species was shown by Vlek *et al.*, 2014. In addition, the number of MSPs per species was found to be significantly correlated with a correct MALDI-TOF MS identification. A higher number of MSPs (commercial BDAL with an additional 510 MSPs) resulted in an increase in correct identifications of approximately 25% (Vlek *et al.*,

2014). Moreover, some recent studies have shown that when very few reference spectra are included in the commercially available database, creation of an in-house library of previously accurately identified reference strains (e.g. determined by rDNA sequencing) improved fast and reliable identification of isolates used to validate or to challenge such spectral library (Stevenson *et al.*, 2010; Cassagne *et al.*, 2011; Posteraro *et al.*, 2012; Firacative *et al.*, 2012, Cassagne *et al.*, 2013; Kolecka *et al.*, 2013; Kolecka *et al.*, 2014).

Examples of MALDI-TOF MS identification of yeasts

A pioneering study explored the application of MALDI-TOF MS using a large panel of yeasts belonging to both *Saccharomycotina* (*Ascomycota*) and *Basidiomycota* (Marklein *et al.* 2009). Eighteen type strains and 267 clinical isolates, 250 of which were *Candida* spp., and the remaining species of *Cryptococcus, Saccharomyces, Trichosporon, Geotrichum, Pichia,* and *Magnusiomyces* spp., were identified by MALDI-TOF MS. The large majority of isolates (92.5%) could be correctly identified. For the remaining 20 isolates, the reference entries had to be made in-house using type strains or molecularly identified isolates and thereafter all isolates could be identified properly.

The taxonomic resolution has been investigated in many studies in which closely related species (i.e. species complexes, sibling species, and hybrids) were involved (Fig. 3.1). In general, no major errors, such as genus misidentifications, were reported in many MALDI-TOF MS-based studies on yeasts (Marklein *et al.* 2009; Stevenson *et al.*, 2010; Cassagne *et al.*, 2013, Kolecka *et al.*, 2013, 2014). Most yeasts can be easily processed and correctly

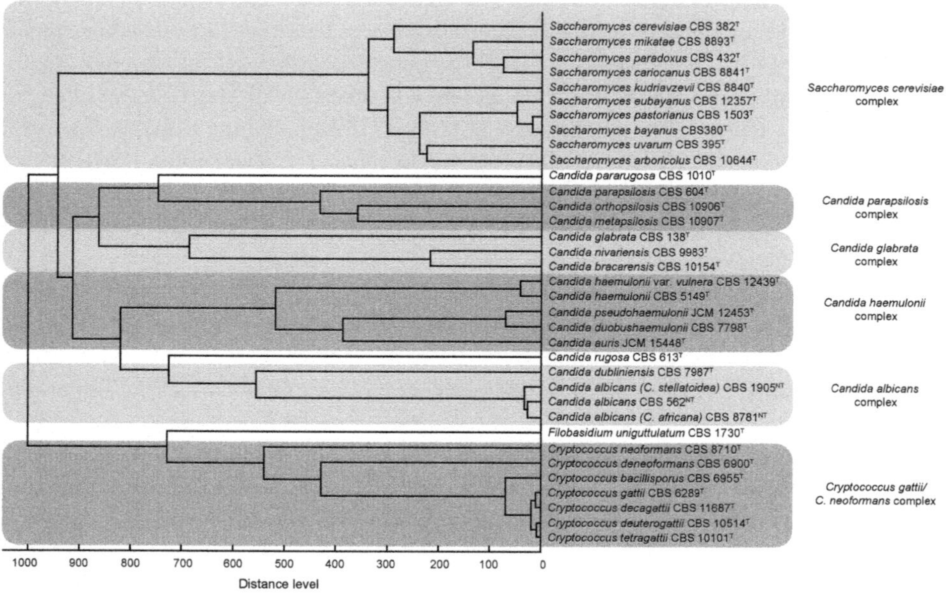

Figure 3.1 Dendrograms representing MALDI-TOF MS potential for a conclusive separation of 'cryptic species' in yeasts; in the ascomycetous genus *Candida* and *Saccharomyces*, and closely related species of the basidiomycetous representatives of *Cryptococcus gatti*/ *Cryptococcus neoformans*.

identified as even sibling species that cannot be distinguished with common biochemical methods can be discriminated by MALDI-TOF MS (Cendejas-Bueno *et al.*, 2012; Firacative *et al.*, 2012; Hagen *et al.*, 2015). However, some users of the MALDI Biotyper (Bruker Daltonics, Germany) system still experience difficulties with the correct identification of uncommon and emerging fungal species as not reliable identifications (log-score values < 1.7) or identifications at the genus level only occurred and some species were not present in the database (Stevenson *et al.*, 2010; van Veen *et al.*, 2010; Bader *et al.*, 2011; Vlek *et al.*, 2014; Riat *et al.*, 2015).

Candida species are ubiquitous organisms that represent the most common fungal human pathogens, viz. *C. albicans, C. glabrata, C. parapsilosis, C. tropicalis* and *C. krusei* (also known as *Pichia kudriavzevii* or *Issatchenkia orientalis*), *C. lusitaniae* and *C. dubliniensis* that may be implicated in infections. The genus *Candida*, however, is not phylogenetically circumscribed and includes all anamorphic (asexual) yeasts belonging to *Saccharomycotina* (Lachance *et al.*, 2011). Several *Candida* species that are not phylogenetically related to the major pathogens can also cause infection, and an increasing number of diverse *Candida* species may be implicated in rare infections (Arendrup *et al.*, 2014). Some of these species differ in their susceptibility to antifungals (Arendrup *et al.*, 2014; Pfaller *et al.*, 2014), hence, a fast and reliable identification of the yeast pathogen at stake is highly important. Given the broad phylogenetic divergence of the potential *Candida* pathogens the database should cover this diversity. A set of clinical *Candida* isolates ($n = 201$) from Hamad Medical Corporation hospital (Doha, Qatar) was identified by MALDI-TOF MS, and biochemical and morphological features, namely Vitek-2 Compact yeast identification system (bioMérieux, France), API ID 32C (bioMérieux, France), and CHROMagar *Candida* plates (Taj-Aldeen *et al.*, 2013). When results of both methods were concordant the isolate was identified, but in case of discordant results species identification was based on a subsequent sequence analysis of rDNA barcode sequences (D1/D2 domains and ITS1/ITS2 regions). All 133 out of 201 isolates belonging to non-*albicans Candida* species, including *C. orthopsilosis* and *C. pararugosa*, were correctly identified by MALDI-TOF MS. Of note is that isolates of *C. orthopsilosis, Meyerozyma guilliermondii* (= *C. guilliermondii*), *C. pararugosa, Kluyveromyces lactis* (= *C. kefyr*), *Cyberlindnera fabiani* (= *C. fabianii*), and *Lodderomyces elongisporus* were incorrectly identified by phenotypic methods resulting in approximately 11% misidentified isolates (Taj-Aldeen *et al.*, 2013).

In contrast to non-molecular methods, the use of nucleic acid sequencing and/or MALDI-TOF MS method allows discrimination between 'cryptic' species. The *C. albicans* complex comprises four species, of which only three are recognized as separate species, namely *C. albicans* (Lachance *et al.*, 2011), *C. africana* (Tietz *et al.*, 1995; Tietz *et al.*, 2001; Romeo and Criseo, 2009) and *C. dubliniensis* (Sullivan *et al.*, 1995), whereas *C. stellatoidea* is considered conspecific with *C. albicans* (Lachance *et al.*, 2011). Although previously discrimination between *C. albicans* and *C. dubliniensis* remained problematic with reliable interpretation of the final test results obtained by the phenotypic methods, such discrimination, however, can be resolved without difficulty by MALDI-TOF MS (Marklein *et al.*, 2009; Bader *et al.*, 2011; Lacroix *et al.*, 2014). Previous examination of atypical *C. albicans* strains and application of molecular methods, e.g. *HWP1* PCR (Romeo and Criseo, 2008), pyrosequencing of ITS2 of rDNA (Borman *et al.*, 2013) improved the discrimination between cryptic species *C. albicans, C. dubliniensis* and *C. africana*. Presently, the identification of *C. africana* with MALDI-TOF MS is problematic (Rosenvinge *et al.*, 2013; A. Kolecka, unpublished observations), but this may be solved when more strains are being added to the database. *C. bracarensis* and *C. nivariensis*, two emerging pathogens that are closely related to *C. glabrata*, constitute another species complex (Lachance *et al.*, 2011; Gabaldon *et al.*, 2013), but they can be distinguished by MALDI-TOF MS (Bader *et al.*, 2011; Santos *et al.*, 2011). Furthermore, MALDI-TOF MS can identify the related *L. elongisporus* and the 'cryptic' species of the *Candida parapsilosis* complex (Hendrickx *et al.*, 2011; Kubesova *et al.*, 2012; Quiles-Melero *et al.*, 2012; Taj-Aldeen *et al.*, 2013; De Carolis *et al.*, 2014a).

The *C. haemulonii* species complex was recently revised and now comprises *C. haemulonii* var. *haemulonii*, *C. haemulonii* var. *vulnera*, *C. duobushaemulonii*, *Candida pseudohaemulonii* and *Candida auris* (Sugita *et al.*, 2006; Cendejas-Bueno *et al.* 2013). Isolates belonging to this complex represent multiresistant yeasts that cannot be easily differentiated by current commercial methods, but ITS rDNA sequencing (Chowdhary *et al.*, 2013; Ramos *et al.*, 2015) and MALDI-TOF MS methods provide conclusive identifications, with the exception of *C. haemulonii* var. *vulnera* that cannot be separated from the typical variety (Cendejas-Bueno *et al.* 2013).

Cryptococcus neoformans (Sanfelice) Vuillemin is an important basidiomycetous pathogen that globally causes approximately 1 million infections with over 600,000 casualties annually. The *Cryptococcus neoformans* species complex has recently been revised and contains at least seven species (Hagen *et al.*, 2015) and MALDI-TOF MS has been successfully applied to identify these *Cryptococcus* species, including some hybrids (McTaggart *et al.*, 2011; Firacative *et al.*, 2012; Posteraro *et al.*, 2012, Hagen *et al.*, 2015). For the latter study an in house database was made comprising 150 MSPs of 88 molecularly well-characterized isolates and this was tested by identifying 424 isolates obtained during previous investigations (Hagen *et al.* 2015). Major errors, such as incorrect genus ID, were not observed and misidentifications occurred only with part of the serotype AD hybrids that, for about 50% of them, were not identified as hybrids but as belonging to one of the parental species that made up the hybrid (Hagen *et al.* 2015).

Isolates belonging to the arthroconidia-forming yeast-like genera *Geotrichum*, *Galactomyces*, *Magnusiomyces*, *Saprochaete*, and *Trichosporon* could be identified by MALDI-TOF MS (Kolecka *et al.*, 2013), but only after inclusion of a large number of species to the reference database (Kolecka *et al.*, 2013). Phenotypic identification of isolates of these two arthroconidia-forming yeast genera is challenging. Importantly, a correct identification, even only at the generic level may contribute to start of correct treatment as *Geotrichum* spp. are azole resistant while *Trichosporon* spp. are resistant to echinocandins (Taj-Aldeen *et al.*, 2009). In *Trichosporon*, separation of isolates belonging to closely related sibling species, *T. inkin*/*T. ovoides*, *T. japonicum*/*T. asteroides* (all four belonging to the ovoides clade) and *T. dermatis*/*T. mucoides* (from the cutaneum clade) was not always possible (Kolecka *et al.* 2013) and may be improved by adding further isolates to the database.

Malassezia species are lipid-dependent or lipophilic basidiomycetous yeasts that occur widely on human skin where they can be involved in skin disorders, such as pityriasis versicolor, folliculitis, seborrhoeic dermatitis (including dandruff), atopic dermatitis and psoriasis (Boekhout *et al.*, 2010; Gaitanis *et al.*, 2012). They can cause rare systemic infections in neonatal and paediatric patients, but also in immunocompromised adult patients with central venous catheters (CVC) through which they receive lipid supplementation (Schleman *et al.*, 2000; Chryssanthou *et al.*, 2001; Nguyen *et al.*, 2001; Kessler *et al.*, 2002; Curvale-Fauchet *et al.*, 2004; Devlin, 2006; Oliveri *et al.*, 2011). Phenotypic identification of *Malassezia* isolates is difficult as it requires specific lipid-supplemented growth media [e.g. modified Dixon's agar, (mDA) or modified Leeming and Notman agar (MLNA)] (Boekhout *et al.*, 2010; Gaitanis *et al.*, 2012), and testing growth requirements using various polysorbate surfactants (i.e. Tweens), Castor oil, growth at different temperatures, and these phenotypic methods give an error rate of 14% when compared to molecular identifications (Boekhout *et al.*, 2010; Gupta *et al.*, 2004). Several hundreds of isolates obtained from skin in various parts of the world could be identified at the species level using MALDI-TOF MS, with ID's concordant with ITS and/or D1/D2 sequencing data (Iatta *et al.*, 2014, Kolecka *et al.*, 2014, Yamamoto *et al.*, 2014). Major errors defined as incorrect genus ID were not observed with log-scores > 1.7 and minor errors, namely misidentification at the species level, were not found either. The cut-off value for correct species identification of *Malassezia* isolates was suggested to be lowered to log-scores > 1.7, as it was already suggested for other yeasts (see above, van Herendael *et al.*, 2012, Kolecka *et al.*, 2013, Vlek *et al.*, 2014).

Recent investigations have also shown an increased occurrence of *Saccharomyces cerevisiae*, a major food yeast, in clinical materials recovered from patients with a weakened immune system. As a rare human pathogen, the species is causing an emerging number of superficial infections and systemic diseases in immunocompromised patients (McCullough *et al.*, 1998; Riquelme *et al.*, 2003; Cassone *et al.*, 2003; Muñoz *et al.*, 2005; de Llanos *et al.*, 2006; Arendrup *et al.*, 2014). The first multicentre study regarding yeast infections in Romania reported MALDI-TOF MS identifications of 11 *S. cerevisiae* isolates that showed decreased fluconazole susceptibility (Minea *et al.*, 2015). The application of MALDI-TOF MS into current microbiological practice allowed correct identification of *S. cerevisiae* isolates in many hospitals worldwide (Bader *et al.*, 2011; Eddouzi *et al.*, 2013; Rosenvinge *et al.*, 2013) and although the *in vitro* susceptibility pattern is similar to that of *C. glabrata*, i.e. with elevated azole MICs (Arendrup *et al.*, 2014), both species can easily be distinguished with this method.

The 'black yeasts' belong to a group of melanized fungi (yeasts and moulds) with a temporary yeast morph may be occasionally involved in human infections that range from cutaneous, pulmonary colonization to systemic or disseminated infection (Chowdhary *et al.*, 2014a,b). Most clinically relevant species belong to genera such as *Exophiala, Cladophialophora, Coniosporium, Cyphellophora, Fonsecaea, Phialophora* and *Rhinocladiella* (Revankar and Sutton, 2010; Seyedmousavi *et al.*, 2013; Seyedmousavi *et al.*, 2014; de Hoog *et al.*, 2000). Other important opportunists are represented by the saprophytic dematiaceous yeast *Aureobasidium pullulans* species complex and the polymorphic, yeast-like fungus *Hortaea werneckii* (previously known as *Exophiala werneckii* or *Cladosporium werneckii*) (Chowdhary *et al.*, 2014b). MALDI-TOF MS-based differentiation and identification was successful for *Exophiala dermatitidis* isolates obtained from the respiratory tracts of patients with cystic fibrosis (Kondori *et al.*, 2015) and *Aureobasidium pullulans* isolated from the bio-aerosols in a hospital in South Africa (Setlhare *et al.*, 2014). In case of *Aureobasidium pullulans* reliable measurements were observed when young (24–48 hours old) cream-like to pink-like colonies are processed (A. Kolecka, unpublished observations). The presence of dark fungal pigments, such as melanin, in the black mould *Aspergillus niger*, resulted in poor statistical discrimination of mass spectral peaks yielding poor or no $[M+H^+]$ ion signals (Buskirk *et al.*, 2011; Hettick *et al.*, 2008b). Identification of pink yeasts, producing carotenoids, belonging to *Rhodotorula* and *Rhodosporidium*, could also be well identified independent on the pigment production, but the number of correct identifications remains limited due to insufficient species coverage in the commercially available database (Pavlovic *et al.*, 2014; A. Kolecka, unpublished observations). Problematic identifications of pink yeasts could also result from poor extraction, especially when the yeast is growing as a slimy culture (A. Kolecka, unpublished observations), and this may be resolved by longer incubation with FA.

Blood cultures/BSI/sepsis

The precise identification of any microorganisms isolated from a positive blood culture is crucial for a life-saving management of the patient. Bloodstream infections (BSIs) and endocarditis represents severe diseases with high mortality and morbidity, especially when patients are treated with inappropriate therapy. Not only bacteria, but also fungi are recognized as causative agents of BSIs. Polymicrobial (mixed) infections occur as well, but, fortunately, such infections occur only rarely (Spanu *et al.*, 2012). Yeasts most commonly isolated from blood cultures are *C. albicans, C. glabrata, C. parapsilosis, C. tropicalis* and *C. krusei* (= *Pichia kudriavzevii*) that account for 90% of cases (Marinach-Patrice *et al.*, 2010). The incidence of candidaemia is growing worldwide and *Candida* infections are an important cause of morbidity and mortality especially in immunocompromised patients, those with hemato-oncological malignancies or individuals with prolonged hospitalization receiving advanced medical treatment (Arendrup, 2010). As long as the identification of yeasts is performed from a solid medium subculture, MALDI-TOF MS application is reported as a superior identification method to the routine phenotypic or biochemical methods (Marinach-Patrice *et al.*, 2010; Taj-Aldeen *et al.*, 2013; Minea *et al.*, 2015; Kolecka *et al.*, 2014). The MALDI-TOF MS identification of yeast isolates directly from blood cultures remains challenging (Ferreira *et al.*, 2011; Ferroni *et al.*, 2010; Yan *et al.*, 2011; Spanu *et al.*, 2012). The human blood is rich in blood cells (such as erythrocytes and leucocytes), proteins, peptides (e.g. haemoglobin or albumin) and other substances that may interfere with the peak pattern of the mass spectrum and, therefore, would complicate the identification process (Marinach-Patrice *et al.*, 2010; Bader, 2013) and result in lower identification rates (Buchan *et al.*, 2012; Ferreira *et al.*, 2011; Yan *et al.*, 2011). Several protocols have been developed to remove human blood components. Steps such as filtration, washing, centrifugation, and saponification were tested to increase the analyte purity and/or enrich the sample before identification with the ethanol/formic acid extraction method. The pretreatment step can also be performed with a commercial kit (Sepsityper) available from Bruker Daltonics (Yan *et al.*, 2011; Schieffer *et al.*, 2014; Bidart *et al.*, 2015). The complete removal of blood cells by a lysis solution is essential for obtaining a high accuracy of the identifications ranging from 91.3% to 100% (Yan *et al.*, 2011; Ferreira *et al.*, 2011; Spanu *et al.*, 2012; Buchan *et al.*, 2012;). In summary, the data available suggest that the MALDI-TOF MS method for analysis of blood cultures is promising but requires further optimization of pretreatment steps.

Typing

MALDI-TOF MS can identify strains of fungal species, but one has also attempted to differentiate strains at the subspecies and variety levels (Pulcrano *et al.*, 2012; Bader, 2013; Dhieb *et al.*, 2015). A further application of MALDI-TOF MS is the typing of isolates, e.g. in nosocomial outbreaks. If possible, it will allow epidemiological analysis of isolates obtained from patients and environmetal samples, but also in controling food fermentations and other biotechnological processes. From the clinical and biotechnological perspectives, microbial typing is challenging as microbes continue to change at the genetic level. Discrete differences between the protein spectra among individual strains will be used for typing strains. As failure in microbial identification by MALDI-TOF MS usually results from an insufficient number of mass peaks due to improper sample preparation, improvement of the number of well resolved peaks in the spectra that can be measured with high reproducibility may yield improved typing resolution (Bader, 2013). MALDI-TOF MS has recently being exploited for the ability to analyse proteomic differences between strains. Using 38 isolates of *Trichosporon asahii* from Greece that were previously correctly identified with score values > 2.0 and by sequence analysis of the LSU and InterGenic Spacer (IGS) rDNA regions, a principal components analysis (PCA) dendrogram of the mass spectra (not based on MSPs) was created in order to test the discriminatory power of MALDI-TOF MS (Kolecka *et al.*, 2013). The resulting four clusters remained separated in the dendrogram, but they did not correspond with the IGS subgroups. Biofilm-positive and biofilm-negative strains of *C. parapsilosis* and *C. metapsilosis* could be distinguished (Kubesova *et al.*, 2012). Application of MALDI-TOF MS for fingerprinting blood isolates of *Candida parapsilosis* sensu stricto from neonatal intensive care units discriminated strains with a resolution that was concordant with that obtained by microsatellite analysis (Pulcrano *et al.*, 2012). Intraspecific characterization of isolates of the *C. parapsilosis* complex was assessed by a comparison of amplified fragment length polymorphism (AFLP) and MALDI-TOF MS-based data (De Carolis *et al.*, 2014a). Here, AFLP showed a better discriminatory power for biotyping strains of all three species *C. parapsilosis*, *C. orthopsilosis* and *C. metapsilosis*, when compared to MALDI-TOF MS data that were highly distinctive at the species level. A recent study, however, did not observe significant concordance between clusters of isolates based on microsatellite length polymorphisms of 33 isolates of *C. albicans* and the clusters in PCA dendrograms of mass spectra data generated by MALDI-TOF MS (Dhieb *et al.*, 2015). Data presented in recent studies seem to strongly support the possibility that MALDI-TOF MS may have potential to be used as a biotyping tool. The MALDI-TOF MS biotyping ability, in comparison with molecular approaches, such as PCR using delta primers and contour-clamped homogenous electric field electrophoresis (CHEF), was recently used for intraspecific discrimination of laboratory and commercial *Saccharomyces cerevisiae* strains into specific subgroups (biotypes) (Moothoo-Padayachie *et al.*, 2013; Usbeck *et al.*, 2014). Such biotyping would be of industrial importance for detection of contaminants or protection of patented strains to ensure product quality and biosafety. An in-house reference database of mass spectra of 44 previously biotyped *S. cerevisiae* strains, commonly employed in local fermentation-based industries, such as wine making, brewing, baking, was created in South Africa (Moothoo-Padayachie *et al.*, 2013). A PCA-based dendrogram for those 44 reference spectra showed five main clusters. The wine yeasts formed a major group, probably due to a common origin as most commercially available strains originated from clonally selected wine yeast of variants or modified hybrid strains. The accuracy of *S. cerevisiae* biotyping was evaluated with two sets of strains; set of 20 blind-coded *S. cerevisiae* strains obtained from an independent research facility and seven blind-coded *S. cerevisiae* strains from a local industrial ethanol-producing plant. All strains were correctly identified at the species level and 18 out of 20 strains matched at the strain level. The two remaining strains turned out to represent problematic contaminants in beer production. For the second subset, only four out of seven strains were correctly recognized at the strain level, and the remaining three incorrectly biotyped strains were recovered from fermentation tanks with low ethanol production; thus MALDI-TOF MS typing showed they were distinct (Moothoo-Padayachie *et al.*, 2013). In modern winemaking quality control of starter cultures and detection of contaminant strains requires the ability of differentiation at both the species and strain levels. The biotyping discriminatory power of MALDI-TOF MS was also tested on a set of 33 *S. cerevisiae* strains from varying wine fermentations and compared with PCR using delta primers (Usbeck *et al.*, 2014). The cluster analysis based on both methods was further investigated whether strain and ecotype-level differentiation is possible, and resulted in a separation of five groups clustering *S. uvarum* (cluster 1), cold wine strains of *S. kudriavzevii* (cluster 2), red wine strains of *S. cerevisiae* (cluster 3), the distillery/wine group of *S. cerevisiae* strains (cluster 4), and a separate cluster 5 for the cold white wine strain Lalvine W15 (Usbeck *et al.*, 2014).

While the use of MALDI-TOF MS for strain typing is still in its infancy, we recommend to follow the strategies for MALDI-TOF MS-based typing as suggested after a comparison was made with the data generated by Pulsed Field Gel Electrophoresis (PFGE) for bacteria (Spinali *et al.*, 2015), and that also may be useful when tying fungal isolates.

1. **Technological and biological mass spectrometry issues.** Analytical errors, such as presence versus absence of mass peaks and variations in their intensity, reduce noise in spectra, consider biological and technical variation, analyte–matrix interaction, selection of the proper matrix solution and use of discrete peaks.
2. **Clonality (reproduction) versus correlation (statistics).** It is difficult to correlate the presence of peaks in mass spectra with phenotypes, such as virulence or resistance as MALDI-TOF MS addresses whole cell [mainly intracellular] proteins, and cell-wall bound proteins. Further bio-informatics analysis may enhance the quality of the analyses.
3. **Sensitivity, specificity and strain sets.** When only a limited number of isolates from a small set of clonal complexes is available this by definition does not capture the entire variation in the total population.
4. **Specific peaks definition and limitation.** Avoid erroneous interpretation of mass spectra, as unstable peaks may be expressed under specific cultivation conditions. Type-specific biomarkers should be recorded consistently with reasonable signal intensities based on biological and analytical replica repetition.
5. **Statistical analysis.** Statistical analysis of peak patterns is important for the discovery and evaluation of type-specific biomarkers or peak patterns.
6. **Primary MALDI-TOF MS identification limitations.** Monitoring of identification problems that may be related to sample preparation, machine alignment and quality of chemicals.

Optimization of this guideline in the laboratory practice and further development of bio-informatics tools will be important for improving future typing studies using MALDI-TOF MS technology in hospital infection control, surveillance programmes of microbial outbreaks, biodefence, quality control of biotechnological production processes, and quarantine regulation maintenance.

Food-borne yeasts

Yeasts of importance to the food industry belong mainly to the ascomycetous genera and less to basidiomycetous genera. Products from alcoholic fermentation (wine, beer), fermented milk products, fermented meat, and baking products largely depend on yeasts. However, yeasts may also be responsible for spoilage of foods and beverages, particularly by osmotolerant – and preservative resistant species, such as *Zygosaccharomyces bailii* (Boekhout and Robert, 2003). In order to prevent or reduce the product spoilage by yeasts, it is essential to detect their presence early and accurately identify the, preferably at the species level. In general in food mycology, accurate identification will help to establish whether the contaminant species represents a threat to the food product or consumer, e.g. due to production of toxic metabolites, such as mycotoxins that are produced by e.g. species of *Aspergillus* and *Fusarium*. Yeasts occurring in man-made environments and with applied importance can be identified and distinguished by phenotypic and biochemical microbiological methods that are time-consuming and not always conclusive. Molecular sequence analysis and MALDI-TOF MS-based identifications however, remain the most reliable methods. The European Food Safety Authority is regularly updating the list of qualified presumption of safety (QPS) biological agents (EFSA Panel on Biological Hazards, 2013). One of the agents, *Saccharomyces cerevisiae*, is considered as a non-pathogenic species, considered as the safest microorganism known and has a GRAS (organism Generally Recognized As Safe) and QPS status, despite that isolates are occasionally isolates from patients. The genus *Saccharomyces* sensu stricto contains *S. arboricola*, *S. bayanus*, *S. cariocanus*, *S. cerevisiae*, *S. eubayanus*, *S. kudriavzevii*, *S. mikate*, *S. paradoxus*, and *S. pastorianus* species (Vaughan-Martini and Martini, 2011) that were distinct in a PCA dendrogram clustering reference mass spectra derived from MALDI-TOF MS (Blattel *et al.*, 2013). An attempt to identify the yeast species involved in the fermentation of 'chicha de jora', the traditional beverage from Peru, was successful with MALDI-TOF MS (Vallejo *et al.*, 2013). Strains present in the final step of fermentation of this drink were identified as *S. cerevisiae* by MALDI-TOF MS and was concordant with identifications based on sequence analysis of D1/D2 domains of the LSU region of rDNA.

A set of 96 randomly selected food-borne yeast isolates, comprising 33 species, was identified using MALDI-TOF MS (Pavlovic *et al.*, 2014). The lack or an inadequate number of reference spectra in the commercial database resulted in ambiguous identifications for 10 out of 33 strains. MALDI-TOF MS scores for these strains ranged between 1.7–1.999 and were presented as inconclusive or discordant results when compared with ITS sequencing and biochemical tests. Out of these 10 strains, the most problematic species belonged to the pink yeasts (*Rhodotorula* sp. *Rhodosporidium* sp., *Sporobolomyces* sp.), *Kazachstania servazzii*, and rare *Candida*, *Cryptococcus* and *Trichosporon* species (Pavlovic *et al.*, 2014). The study of Usbeck *et al.*, (2013) showed that optimized sample preparation and measurement parameterization of the MALDI-TOF MS method resulted in a reliable differentiation between three beverage spoiling yeasts, namely *Saccharomyces cerevisiae* (cited as *Saccharomyces cerevisiae* var. *diastaticus*), *Wickerhamomyces anomalus and Debaryomyces hansenii*. Note that based on the current taxonomic names of yeasts *S. cerevisiae* var. *diastaticus* is a synonym of *S. cerevisiae* (Vaughan-Martini and Martini, 2011). Usbeck *et al.* (2013) also recommend using the ETOH/FA extraction method as it enabled the highest reproducible peak number and an initial laser intensity that was set at 50–60%.

Analysis of filamentous fungi

In the modern clinical laboratory, time of identification of colony-growing filamentous fungi (i.e. moulds) can be shortened through the use of a purely biophysical method such as MALDI-TOF MS (Clark *et al.*, 2013; Sanguinetti and Posteraro, 2014). In contrast to its rapid expansion in clinical bacteriology, MALDI-TOF MS slowly gained acceptance in clinical mycology, especially with respect to filamentous fungi. This was because of the biological complexity of these organisms that accounts for the coexistence of different phenotypes (i.e. hyphal or conidial) in the same fungal isolate (Santos *et al.*, 2010), as well as the presence of a robust cell wall that represents an obstacle for the direct identification of fungi using a basic sample preparation such as an 'intact cell' (IC) or 'whole cell' MS methods (Posteraro *et al.*, 2013).

In the literature separation exists between studies that used (i) an IC approach, in which a single colony is smeared directly onto a MALDI target plate and covered by an acidic organic MALDI matrix solution, and studies that used (ii) a cell lysis (CL) approach, in which an ethanol–formic acid procedure allows complete protein extraction, that consists of a short incubation and centrifugation steps prior to depositing the supernatant onto the target plate. Of these two methodological approaches, the former, possibly including a short on-target extraction with a formic acid solution, is recommended for use with the Saramis, Andromas and Vitek MS systems, whereas the latter is superior with the Bruker MALDI Biotyper (Vermeulen *et al.*, 2012). Mould isolates in these studies were identified by comparing their own spectra with those included in an in-house made reference library/database, and it was observed that identification failures, mainly no identifications, occurred with fungal species for which no specific entries were present in the reference databases employed (Vermeulen *et al.*, 2012). It should be noted that inadequate MALDI-TOF MS results are often due to discrepancies between the methods used for clinical isolate testing and spectral database construction (Bader, 2013), whereas the increase of the number of mass spectra generated from distinct subcultures of fungal isolates, for each species represented in the reference library, may enhance the accuracy of the MALDI-TOF MS-based identification of moulds (Normand *et al.*, 2013).

Sample preparation methods

Starting from the initial descriptions of the IC MALDI-TOF MS analysis, in which mass spectrometric profiles are acquired by desorption of specific peptide/protein biomarkers from the cell/spore surface of a given microorganism (Fenselau and Demirev, 2001), several efforts have been made to develop sample preparation methods, together with sample deposition techniques, to be employed for the MALDI-TOF MS identification of filamentous fungi (Chalupová *et al.*, 2014). In their first versions, these methods also rely on the pre-extraction of proteins by acidified solvents (Welham *et al.*, 2000) or by the help of a bead beating step prior to the MS analysis (Hettick *et al.*, 2008a; 2008b). For identification/characterization of clinically relevant fungi of *Aspergillus, Penicillium, Fusarium,* or representatives of zygomycetes and dermatophytes (Clark *et al.*, 2013; Sanguinetti and Posteraro, 2014), the following sample preparation methods are available: (i) direct colony deposition ('toothpick method') methods, that consist in spotting of a part of a colony without prior extraction or lysis; and (ii) extracted-colony deposition methods, that consist in the spotting of a colony-derived extract after chemical or mechanical disruption of fungal cells.

In the first group the procedure involves the following steps (Bille *et al.*, 2012; Iriart *et al.*, 2012):

- a Transfer of a small piece of fungal colony grown on Sabouraud dextrose agar onto the MALDI target plate.
- b Overlay with a ready-to-use HCCA matrix.
- c Dry the fungal spot completely.
- d Proceed with the MALDI-TOF MS analysis.

In the second group the procedure involves the following steps (Cassagne *et al.*, 2011; Lau *et al.*, 2013):

- a Harvest of a small sample, i.e. fungal hyphae and spores, by scraping the fungal colony grown on Sabouraud dextrose agar.
- b Mix the fungal sample in sterile water with absolute ethanol.
- c Centrifugation followed by incubation of the pellet in 70% formic acid.
- d Add 100% acetonitrile and incubate.
- e Centrifugation, followed by recovery of the supernatant.
- f Deposition of a drop of supernatant on a spot of the MALDI target plate.
- g Proceed as in steps b to d of the procedure described above.

Alternatively, steps (a) and (b) can be replaced by a mechanical lysis steps as follows:

- a Place a small piece of fungal isolate in absolute ethanol mixed with zirconium–silica beads.
- b Emulsify the fungal suspension thoroughly.
- c Vortex at maximum speed and spin down.
- d Harvesting the pellet to be treated as in step (c), and so on.

Compared with the direct-spotting method, the protein-extraction method is time-consuming, because the fungal colonies must be suspended in Eppendorf tubes, mixed with solvents, and centrifuged several times. Although simplified and promising procedures began to appear since the year 2011 (Alanio *et al.*, 2011; De Carolis *et al.*, 2012a), it has been questioned whether the use of protein extraction for filamentous fungi would result in better quality spectra with enhanced sensitivity and specificity, presumably because cleaner spectra are produced without interference of cell constituents (Lau *et al.*, 2013). It is also questionable whether heterogeneity of mould colonies grown on solid media would be higher than when moulds are grown in liquid cultures. In an attempt to minimize the effect of culture conditions and to facilitate the production of uniform mycelium, Bruker Daltonics launched in 2012 a separate additional library (Fungi library v.1.0) for the identification of filamentous fungi grown in liquid media (Posteraro *et al.*, 2013).

Nevertheless, we daily experience that a simple and fast protocol, by which a drop of a water mixture of mycelium and/or conidia is directly spotted on the target plate prior to the matrix application for MALDI-TOF MS analysis (De Carolis *et al.*, 2012a), is able to provide accurate identification of mould isolates cultured from clinical specimens (Fig. 3.2). In the routine laboratory practice, irrespective of which method(s) is chosen, clinical mycologists should take into consideration to use a biosafety cabinet during sample processing in order to minimize the risk of spore aerosolization with potential laboratory-acquired infection or contamination.

Databases

Several authors, including us (BP, MS), have demonstrated that use of an updated commercial spectral database or a carefully engineered custom database is crucial to achieve reliable results. Low identification rates can be expected for real samples, such as clinical mould isolates, when few less-common or uncommonly occurring fungal species are included in the commercial databases (Patel, 2015). To date, two extensive clinical databases have been developed for the identification of moulds by MALDI-TOF MS. One database was constructed to comprise 152 species of moulds by using reference spectra of 294 isolates, of which 58 reference strains were obtained from well-maintained collections and 236 isolates originated from patients' specimens in NIH premises (Lau *et al.*, 2013). When challenged with spectra from a set of 421 clinical isolates, the NIH database gave acceptable species-level identification (score ≥ 2.0) for around 90% (370/421) of isolates tested, while the Bruker Biotyper library (BDAL v.3.3.1.0_4110–4613) alone identified only three isolates (0.7%) (Lau *et al.*, 2013). The second database contained 2832 reference spectra covering 347 species of moulds that was constructed using 708 isolates with four spectra each (Gautier *et al.*, 2014). When challenged with spectra from 1107 clinical isolates, primarily from

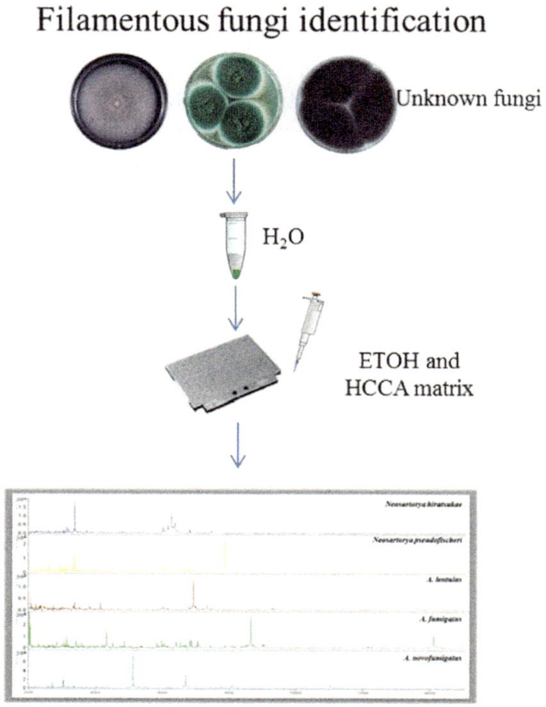

Figure 3.2 Schematic view of the use of MALDI-TOF MS in mycological diagnostics with respect to the fungal identification.

respiratory tract specimens, this database enabled species-level identification for 98.8% (1094/1107) of isolates corresponding to 107 species. Importantly, re-assessment of a set of clinical isolates previously identified using phenotypic methods, including MALDI-TOF MS, allowed obtaining an increase in species diversity from 16 to 42 taxa (Gautier *et al.*, 2014).

As above mentioned, the Bruker Daltonics mould database (Filamentous Fungi Library 1.0) was constructed using 24(48)h-old liquid cultures subjected to complete protein extraction. When this database was evaluated using 83 well-characterized, non-dermatophyte and non-dematiaceous moulds, the authors showed that 78% and 54% of the isolates were identified at the genus and species levels, respectively, and that reducing the species cutoff value to ≥1.7 improved species identification to 71%, without increasing misidentifications (Schulthess *et al.*, 2014). In a prospective evaluation performed on 200 consecutively obtained clinical mould isolates, they were able to achieve genus and species identification rates of 84% and 79%, respectively, using a species cut-off value of 1.7. Importantly, the percentage of not identified isolates was 16.5%, opposed to 4% of no identifications that occurred when the conventional identification algorithm was employed (Schulthess *et al.*, 2014).

Examples of identification of moulds by MALDI-TOF MS

MALDI-TOF MS has been applied for species characterization of clinically relevant moulds, that was performed through direct surface analysis of fungal colonies which were arbitrarily categorized as young and mature on the basis of their size (diameter < 1 cm for young colony; diameter about 3 cm for mature colony) (De Carolis *et al.*, 2012a). For each of 109 culture collection strains, representing 55 species of *Aspergillus* (33 species), *Fusarium* (12 species), and *Mucorales* (10 species), to be included in the reference database, species-specific spectral profiles of young and mature colonies were obtained, using water suspensions of superficial fungal material (mycelium and/or conidia) and the Bruker MALDI Biotyper software.

After construction of an in-house reference database, 103 blind-coded fungal isolates routinely collected in our clinical microbiology laboratory were similarly processed and analysed. Excluding nine isolates that belonged to species not included in our database, we identified 91 (96.8%) out of 94 isolates of *Aspergillus*, *Fusarium*, and *Mucorales*, in agreement to their species names determined by the multilocus sequencing method with score values all > 2.0 that the manufacturer recommended as a cutoff value for reliable species-level identification. Three isolates, namely one *Emericella nidulans*, one *Aspergillus niger* and one *Aspergillus versicolor*, could be identified only at the genus level with score values of 1.817, 1.874, and 1.796, respectively, but, interestingly, their species designation was concordant with that obtained by the sequencing method. In contrast, isolates belonging to species that were not included in the reference database, all had identification score values < 1.7, confirming the specificity of fungal identification by MALDI-TOF MS.

MALDI-TOF MS was able to separate *Aspergillus flavus* and *Aspergillus oryzae* at the species level. This is remarkable as these species are difficult to discriminate by means of β-tubulin sequence analysis, and thus requires the use of a labour-intensive DNA-based technique for their differentiation. Indeed, hierarchical cluster analysis (not PCA) of selected isolates of *Aspergillus* section *Flavi* showed that clinical isolates of *A. oryzae* and *A. flavus* grouped together with their corresponding reference strains but that they formed clusters distinct from those of isolates of *Aspergillus parasiticus* and *Aspergillus alliaceus* (De Carolis *et al.*, 2012a). Also, five of these *A. flavus* isolates included in the analysis could be further distinguished from each other, thus posing the basis to use a MALDI-TOF MS-based phenotypic approach for discriminating non(alfa)toxigenic strains (highly related to *A. oryzae*) from a toxigenic strains of *A. flavus*.

Several studies employing MALDI-TOF MS have recently been dedicated to identification and characterization of dermatophytes belonging to the genera *Trichophyton, Microsporum, Epidermophyton* and *Artroderma* (Theel *et al.*, 2011; Nenoff *et al.*, 2013, Packeu *et al.*, 2013). This group of keratinophilic fungi represents a diagnostic challenge using standard morphological methods, at least for uncommon species, but their identification at the species level is often needed for the appropriate treatment of dermatomycoses. In one of these studies (Nenoff *et al.*, 2013), MALDI-TOF MS analysis following the construction of an in-house database was shown to match identifications obtained with gene sequencing for 283 of 285 (99.3%) dermatophyte isolates. In another study (Packeu *et al.*, 2013), a reference database was constructed with 17 strains of six different species belonging to the *Trichophyton mentagrophytes* complex. When this database was challenged with 54 dermatophyte strains from well-established collections, MALDI-TOF MS was able to discriminate between closely related species of this complex with MALDI-TOF MS data correlating with phylogenetic data.

Use of MALDI-TOF MS was also evaluated for other filamentous fungi, such as species of *Ramularia, Trichoderma* and *Alternaria*. *Trichoderma brevicompactum* was recognized as a distinct species (Degenkolb *et al.* 2008) and it was possible to discriminate between closely related species of *Alternaria* (Brun *et al.*, 2013) and *Ramularia* (Videira *et al.*, 2015). In the later study, an in-house database was created for plant pathogenic *Ramularia* species and MALDI-TOF MS was able to correctly identify all representatives, including *R. glenni* and *R. plurivora* that were obtained from human samples.

Susceptibility testing of yeasts and filamentous fungi

The widespread use of antifungal agents, particularly fluconazole and more recently caspofungin, has in part driven the emergence of fungal species and strains exhibiting reduced susceptibility or resistance to these commonly used antifungal agents. This reinforces the importance of performing antifungal susceptibility testing (AFST) on the isolates recovered from infected patients, as well as the need for reproducible and clinically relevant AFST methods to be utilized as a guide for appropriate antifungal therapy (Posteraro *et al.*, 2014). Two standard broth microdilution techniques, namely that of the Clinical and Laboratory Standards Institute (CLSI) and that of the European Committee on Antimicrobial Susceptibility Testing (EUCAST), are accepted for performing AFST of *Candida* and filamentous fungi. As an alternative approach to these reference techniques, a MALDI-TOF MS-based assay was developed for testing antifungal susceptibilities of clinically relevant *Candida* and *Aspergillus* species (De Carolis *et al.*, 2012b).

Sample preparation methods

For the development of the assay, we took advantage of the previous observation that exposure of *Candida albicans* to fluconazole for 15h induced a modification in the proteomic profile (Marinach *et al.*, 2009). This led to the formulation of a new endpoint, the minimal profile change concentration (MPCC) as opposed to the classical minimum inhibitory concentration (MIC) parameter. The MPCC is expressed as the lowest concentration of antifungal drug, fluconazole in this case, that induces significant changes in the mass spectrum when compared with peak patterns documented in the spectra obtained with the yeast incubated at different drug dilutions (from 128 to 0.125 µg/ml) for 15 hours. Thus, a statistical method is employed to calculate the degree of similarity between the spectra recorded at each of different intermediate fluconazole concentrations and the spectra recorded at each of two extreme concentrations (fluconazole 128 µg/ml and no fluconazole). Accordingly, the spectra of the test organism are classified as 'nearer to the 128 µg/ml' or 'nearer to fluconazole-free' spectrum to define the organism as fluconazole-resistant or susceptible, respectively (Marinach *et al.*, 2009).

This MALDI-TOF MS-based AFST method applies the composite correlation index (CCI) analysis, which is a part of the Bruker Biotyper software, to calculate a correlation matrix that involves the following steps:

a Incubate fungal cells of the test organism (yeast or mould) for 15 hours in the presence of serial antifungal (caspofungin) drug concentrations (including the drug free one), ranging from 0.5 to 64 µg/ml.
b Acquire mass spectra from drug-exposed and drug-free fungal cells.
c Match each concentration's spectrum against the spectra at each of the two extreme concentrations (null or maximal) of the drug.
d Determine the MPPC endpoint, defined as the CCI value at which a spectrum is more similar to the one observed at the maximal (64 µg/ml) drug concentration (maximum CCI) than the spectrum observed with drug free sample (0 µg/ml) (null CCI).

MALDI-TOF MS-based AFST has the advantage of eliminating subjective readouts, which occur with the microdilution broth-based reference methods when filamentous fungi are tested, and it gives better discrimination of isolates with trailing growth. The endpoint readings achievable with the above presented version of our assay result in saving of time, i.e. 15 hours versus 24 hours for the reference methods. For this reason, we propose a simplified version of our assay, namely mass spectrometry-AFST (ms-AFST), which was established to discriminate susceptible and resistant isolates of *Candida albicans* after a short-time incubation with 'breakpoint' concentrations of caspofungin (Vella *et al.*, 2013). This assay, of which a schematic representation is shown in Fig. 3.3, relies on:

a Incubate yeast cells in the presence of three drug concentrations, namely 0, 0.03, or 32 µg/ml of antifungal (caspofungin) for 3 hours.
b Acquire fungal spectra at three drug concentrations.
c Compare the above obtained spectra to create individual CCI matrices.
d Categorize the yeasts as susceptible or resistant to caspofungin if the CCI values of spectra at 0.03 and 32 µg/ml are higher or lower than the CCI values of spectra at 0.03 and 0 µg/ml, respectively.

Examples of susceptibility testing

Using a panel of a wild-type and echinocandin-resistant *FKS* mutants of *Candida* (34 isolates) and *Aspergillus* (10 isolates), we found that the values of MPCC were in agreement

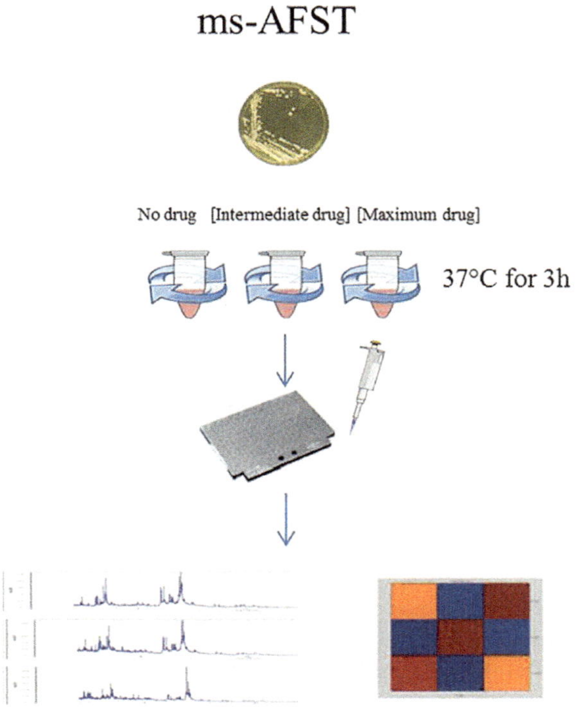

Figure 3.3 Schematic view of the use of MALDI-TOF MS in mycological diagnostics with respect to the antifungal susceptibility testing.

with the values of MIC or minimum effective concentration (MEC) for 100% of the isolates tested (De Carolis *et al.*, 2012b). In contrast, applying the clinical breakpoints (CBPs) values recently proposed for caspofungin yielded a 94.1% categorical agreement with the CLSI reference method. Only two isolates of *Pichia kudriavzevii* (= *Candida krusei*) were misclassified as intermediate (minor errors) by MALDI-TOF MS. All isolates of *Aspergillus fumigatus* and *Aspergillus flavus* showed MPCC values of 0.5 and 0.25 µg/ml, respectively. Importantly, the MPCC values were in accordance with the MECs values (De Carolis *et al.*, 2012b). When testing 65 *C. albicans* isolates, including 13 *FKS1* mutants using the ms-AFST assay, the isolates were classified as susceptible or resistant to caspofungin on the basis of the CCI values of their mass spectra (Vella *et al.*, 2013). According to the *FKS1* genotype, all *C. albicans* isolates ($n = 51$, 100%) were correctly identified as caspofungin-susceptible, whereas 10 (90.9%) out of 11 isolates were correctly identified as caspofungin-resistant. Thus, we found a 98.4% categorical agreement with only one major error, namely for an isolate harbouring a D648Y *FKS1* genotype. This mutation is considered to confer a lower level of echinocandin resistance, thus making it a challenge to identify the isolates possessing such a mutation (Vella *et al.*, 2013). We acknowledged that the ms-AFST, in its present form, obliges laboratory users to utilize two AFST assays (i.e. ms-AFST and the CLSI/EUCAST method), and requires personnel to be trained in using software algorithms, such as CCI-based matching. However, this new rapid AFST method has the potential to be extended to other antifungals or resistance phenotypes, because it seems not to depend on the detection of specific genes or proteins.

Quality control

Microbial identification is an essential component of the work flow in the microbiological laboratory and control of infection services provided worldwide. MALDI-TOF MS is a state-of-the-art method in microbial identification providing reliable identification of diverse pathogens emerging in humans and animals or microbes of environmental or food chain origin. To ensure public health and biosafety through microbial identification, MALDI-TOF MS can be enrolled for the control of microbial hazards associated with many industrial products and could also meet the requirements and principles of overall quality control programmes (Willinger and Haase, 2013; Arendrup *et al.*, 2015). Hospitals, water and food suppliers, food and beverage industries, manufacturers of different biofermentation products as well as microbial culture collections, that need to monitor their production chains on a regular basis, can benefit from this method. Recent reports have shown that the MALDI-TOF MS is an effective monitoring system for use in the culture collections. The identity of closely related members of the *C. parapsilosis* complex as maintained in the BCCM/IHEM collection was resolved in a retrospective study (Hendrickx *et al.*, 2011). A later prospective study on the performance of MALDI-TOF MS in the routine identification of filamentous fungi in the same collection resulted in 84% correct identification (Becker *et al.*, 2014). The same study proposed a workflow dedicated to the identity of preserved strains in culture collection (Becker *et al.*, 2014). Next to the molecular identification of many fungi by molecular barcoding a retrospective identification of many types of yeast in the CBS fungal collection with a workflow combining MALDI-TOF MS was evaluated (e.g. Cendejas-Bueno *et al.*, 2012; Kolecka *et al.*, 2013, 2014; Hagen *et al.*, 2015). It was shown that MALDI-TOF MS gave conclusive identifications concordant to molecular barcoding.

In a multicentre test, established and coordinated by the CBS-KNAW, (Utrecht, the Netherlands) in which the performance of the MALDI Biotyper (Bruker) regarding the identification of yeasts was investigated in a clinical setting (Vlek *et al.*, 2014), considerable differences were observed between laboratories, mainly regarding the sample preparation method, but also some errors were noted that hinted at exchange of strains during the work flow (T. Boekhout, unpublished observations). This demonstrates the need for quality control checks of diagnostic laboratories. Collaboration between scientists working in different countries and settings is therefore essential for further improvement and monitoring of the quality of the MALDI-TOF MS work flow for microbial identification. This could be realized by distribution of well-characterized sets of test strains, e.g. from culture collections, in collaboration with quality maintain organizations or companies.

MALDI-TOF MS will be further optimized in the near future, especially with respect to further improvement of the direct transfer method for both yeasts and moulds, susceptibility testing, epidemiological typing, phenotyping e.g. of pathovars. It is also clear that further automation of the preparatory and analytical procedures may improve efficiency as well as reproducibility of identifications. For all these applications and developments further improvement of bio-informatics tools will be essential.

References

Agustini, B.C., Silva, L.P., Bloch, C.Jr, Bonfim, T.M., and da Silva, G.A. (2014). Evaluation of MALDI-TOF mass spectrometry for identification of environmental yeasts and development of supplementary database. Appl. Microbiol. Biotechnol. *98*, 5645–5654.

Alanio, A., Beretti, J.L., Dauphin, B., Mellado, E., Quesne, G., Lacroix, C., Amara, A., Berche, P., Nassif, X., and Bougnoux, M.E. (2011). Matrix-assisted laser desorption ionization time-of-flight mass spectrometry for fast and accurate identification of clinically relevant *Aspergillus* species. Clin. Microbiol. Infect. *17*, 750–755.

Arendrup, M.C. (2010). Epidemiology of invasive candidiasis. Curr. Opin. Crit. Care. *16*, 445–452.

Arendrup, M.C., Boekhout, T., Akova, M., Meis, J.F., and Cornely, O.A., European Society of Clinical Microbiology and Infectious Diseases Fungal Infection Study Group, and European Confederation of Medical Mycology. (2014). ESCMID and ECMM joint clinical guidelines for the diagnosis and management of rare invasive yeast infections. Clin. Microbiol. Infect. *20* (Suppl. 3), 76–98.

Arendrup, M.C., Posteraro, B., Sanguinetti, M., and Guinea, J. (2015). The state-of-the-art mycology laboratory: visions of the future. Curr. Fungal Infect. Rep. *9*, 37–51.

Bader, O. (2013). MALDI-TOF-MS-based species identification and typing approaches in medical mycology. Proteomics *13*, 788–799.

Bader, O., Weig, M., Taverne-Ghadwal, L., Lugert, R., Gross, U., and Kuhns, M. (2011). Improved clinical laboratory identification of human pathogenic yeasts by matrix-assisted laser desorption ionization time-of-flight mass spectrometry. Clin. Microbiol. Infect. *17*, 1359–1365.

Becker, P.T., de Bel, A., Martiny, D., Ranque, S., Piarroux, R., Cassagne, C., Detandt, M., and Hendrickx, M. (2014). Identification of filamentous fungi isolates by MALDI-TOF mass spectrometry: clinical evaluation of an extended reference spectra library. Med. Mycol. *52*, 826–834.

Bidart, M., Bonnet, I., Hennebique, A., Kherraf, Z.E., Pelloux, H., Berger, F., Cornet, M., Bailly, S., and Maubon, D. (2015). An in-house assay is superior to Sepsityper® for the direct MALDI-TOF mass spectrometry identification of yeast species in blood culture. J. Clin. Microbiol. *53*, 1761–1764.

Bille, E., Dauphin, B., Leto, J., Bougnoux, M.E., Beretti, J.L., Lotz, A., Suarez, S., Meyer, J., Join-Lambert, O., Descamps, P., *et al.* (2012). MALDI-TOF MS Andromas strategy for the routine identification of bacteria, mycobacteria, yeasts, *Aspergillus* spp. and positive blood cultures. Clin. Microbiol. Infect. *18*, 1117–1125.

Blättel, V., Petri, A., Rabenstein, A., Kuever, J., and König, H. (2013). Differentiation of species of the genus *Saccharomyces* using biomolecular fingerprinting methods. Appl. Microbiol. Biotechnol. *97*, 4597–4606.

Boekhout, T., Gueho, E., Mayser, P., and Velegraki, A. (2010). *Malassezia* and the Skin (Springer-Verlag, Berlin, Germany).

Boekhout, T., and Robert, V. (2003). Yeasts in Food: Beneficial and Detrimental Aspects (B. Behr's Verlag GmbH & Co., Hamburg, Germany).

Borman, A.M., Szekely, A., Linton, C.J., Palmer, M.D., Brown, P., and Johnson, E.M. (2013). Epidemiology, antifungal susceptibility, and pathogenicity of *Candida africana* isolates from the United Kingdom. J. Clin. Microbiol. *51*, 967–972.

Brun, S., Madrid, H., Gerrits van der Ende, B., Andersen, B., Marinach-Patrice, C., Mazier, D., and de Hoog, G.S. (2013). Multilocus phylogeny and MALDI-TOF analysis of the plant pathogenic species *Alternaria dauci* and relatives. Fung. Biol. *117*, 32–40.

Buchan, B.W., Riebe, K.M., and Ledeboer, N.A. (2012). Comparison of the MALDI Biotyper system using Sepsityper specimen processing to routine microbiological methods for identification of bacteria from positive blood culture bottles. J. Clin. Microbiol. *50*, 346–352.

Buskirk, A.D., Hettick, J.M., Chipinda, I., Law, B.F., Siegel, P.D., Slaven, J.E., Green, B.J., and Beezhold, D.H. (2011). Fungal pigments inhibit the matrix-assisted laser desorption/ionization time-of-flight mass spectrometry analysis of darkly pigmented fungi. Anal. Biochem. *411*, 122–128.

Cassagne, C., Cella, A.L., Suchon, P., Normand, A.C., Ranque, S., and Piarroux, R. (2013). Evaluation of four pretreatment procedures for MALDI-TOF MS yeast identification in the routine clinical laboratory. Med. Mycol. *51*, 371–377.

Cassagne, C., Ranque, S., Normand, A.C., Fourquet, P., Thiebault, S., Planard, C., Hendrickx, M., and Piarroux, R. (2011). Mould routine identification in the clinical laboratory by matrix-assisted laser desorption ionization time-of-flight mass spectrometry. PLoS One *6*, e28425.

Cassone, M., Serra, P., Mondello, F., Girolamo, A., Scafetti, S., Pistella, E., and Venditti, M. (2003). Outbreak of *Saccharomyces cerevisiae* subtype *boulardii* fungemia in patients neighboring those treated with a probiotic preparation of the organism. J. Clin. Microbiol. *41*, 5340–5343.

Cendejas-Bueno, E., Kolecka, A., Alastruey-Izquierdo, A., Theelen, B., Groenewald, M., Kostrzewa, M., Cuenca-Estrella, M., Gomez-Lopez, A., and Boekhout, T. (2012). Reclassification of the *Candida haemulonii* complex as *Candida haemulonii* (*C. haemulonii* group I), *C. duobushaemulonii* sp. nov. (*C. haemulonii* group II), and *C. haemulonii* var. *vulnera* var. nov.: three multiresistant human pathogenic yeasts. J. Clin. Microbiol. *50*, 3641–3651.

Chalupová, J., Raus, M., Sedlářová, M., and Šebela, M. (2014). Identification of fungal microorganisms by MALDI-TOF mass spectrometry. Biotechnol. Adv. 32, 230–241.

Chowdhary, A., Meis, J.F., Guarro, J., de Hoog, G.S., Kathuria, S., Arendrup, M.C., Arikan-Akdagli, S., Akova, M., Boekhout, T., Caira, M., *et al.* (2014a). ESCMID and ECMM joint clinical guidelines for the diagnosis and management of systemic phaeohyphomycosis: diseases caused by black fungi. Clin. Microbiol. Infect. *20* (Suppl. 3), 47–75.

Chowdhary, A., Perfect, J., and de Hoog, G.S. (2014b). Black Molds and Melanized Yeasts Pathogenic to Humans. Cold Spring Harb. Perspect. Med. *5*, a019570.

Chowdhary, A., Sharma, C., Duggal, S., Agarwal, K., Prakash, A., Singh, P.K., Jain, S., Kathuria, S., Randhawa, H.S., Hagen, F., *et al.* (2013). New clonal strain of *Candida auris*, Delhi, India. Emerg. Infect. Dis. *19*, 1670–1673.

Chryssanthou, E., Broberger, U., and Petrini, B. (2001). *Malassezia pachydermatis* fungaemia in a neonatal intensive care unit. Acta Paediatr. *90*, 323–327.

Clark, A.E., Kaleta, E.J., Arora, A., and Wolk, D.M. (2013). Matrix-assisted laser desorption ionization-time of flight mass spectrometry: a fundamental shift in the routine practice of clinical microbiology. Clin. Microbiol. Rev. *26*, 547–603.

Claydon, M.A., Davey, S.N., Edwards-Jones, V., and Gordon, D.B. (1996). The rapid identification of intact microorganisms using mass spectrometry. Nat. Biotechnol. *14*, 1584–1586.

Curvale-Fauchet, N., Botterel, F., Legrand, P., Guillot, J., and Bretagne, S. (2004). Frequency of intravascular catheter colonization by *Malassezia* spp. in adult patients. Mycoses. 47, 491–494.

De Carolis, E., Posteraro, B., Lass-Flörl, C., Vella, A., Florio, A.R., Torelli, R., Girmenia, C., Colozza, C., Tortorano, A.M., Sanguinetti, M., and Fadda, G. (2012a). Species identification of *Aspergillus*, *Fusarium* and *Mucorales* with direct surface analysis by matrix-assisted laser desorption ionization time-of-flight mass spectrometry. Clin. Microbiol. Infect. *18*, 475–484.

De Carolis, E., Vella, A., Florio, A.R., Posteraro, P., Perlin, D.S., Sanguinetti, M., and Posteraro, B. (2012b). Use of matrix-assisted laser desorption ionization-time of flight mass spectrometry for caspofungin susceptibility testing of *Candida* and *Aspergillus* species. J. Clin. Microbiol. *50*, 2479–2483.

De Carolis, E., Hensgens, L.A., Vella, A., Posteraro, B., Sanguinetti, M., Senesi, S., and Tavanti, A. (2014a). Identification and typing of the *Candida parapsilosis* complex: MALDI-TOF MS vs. AFLP. Med. Mycol. *52*, 123–130.

De Carolis, E., Vella, A., Vaccaro, L., Torelli, R., Posteraro, P., Ricciardi, W., Sanguinetti, M., and Posteraro, B. (2014b). Development and validation of in-house database for MALDI-TOF MS-based yeast identification using a fast protein extraction procedure. J. Clin. Microbiol. *52*, 1453–1458.

de Hoog, G.S., Guarro, J., Gene, J., and Figueras, M.J. (2000). Atlas of Clinical Fungi (Utrecht, Netherlands: Centraalbureau voor Schimmelcultures/Reus, Spain: Universitat Rovira i Virgili).

de Llanos, R., Querol, A., Pemán, J., Gobernado, M., and Fernández-Espinar, M.T. (2006). Food and probiotic strains from the *Saccharomyces cerevisiae* species as a possible origin of human systemic infections. Int. J. Food Microbiol. *110*, 286–290.

Degenkolb, T., Dieckmann, R., Nielsen, K.F., Gräfenhan, T., Theis, C., Zafari, D., Chaverri, P., Ismaiel, A., Bruckner, H., von Dohren, H., *et al.* (2008). The *Trichoderma brevicompactum* clade: a separate lineage with new species, new peptaibiotics, and mycotoxins. Mycological progress 7, 177–219.

Devlin, R.K. (2006). Invasive fungal infections caused by *Candida* and *Malassezia* species in the neonatal intensive care unit. Adv. Neonatal Care *6*, 68–77.

Dhieb, C., Normand, A.C., L'Ollivier, C., Gautier, M., Vranckx, K., El Euch, D., Chaker, E., Hendrickx, M., Dalle, F., Sadfi, N., *et al.* (2015). Comparison of MALDI-TOF mass spectra with microsatellite length polymorphisms in *Candida albicans*. J. Mass Spectrom. *50*, 371–377.

Dhiman, N., Hall, L., Wohlfiel, S.L., Buckwalter, S.P., and Wengenack, N.L. (2011). Performance and cost analysis of matrix-assisted laser desorption ionization-time of flight mass spectrometry for routine identification of yeast. J. Clin. Microbiol. *49*, 1614–1616.

Eddouzi, J., Hofstetter, V., Groenewald, M., Manai, M., and Sanglard, D. (2013). Characterization of a new clinical yeast species, *Candida tunisiensis* sp. nov., isolated from a strain collection from Tunisian hospitals. J. Clin. Microbiol. *51*, 31–39.

EFSA BIOHAZ Panel (EFSA Panel on Biological Hazards) (2013). Scientific opinion on the maintenance of the list of QPS biological agents intentionally added to food and feed (2013 update). EFSA Journal *11*, 3449.

Fell, J.W., Boekhout, T., Fonseca, A., Scorzetti, G., and Statzell-Tallman, A. (2000). Biodiversity and systematics of basidiomycetous yeasts as determined by large-subunit rDNA D1/D2 domain sequence analysis. Int. J. Syst. Evol. Microbiol. *50*, 1351–1371.

Fenselau, C., and Demirev, P.A. (2001). Characterization of intact microorganisms by MALDI mass spectrometry. Mass Spectrom. Rev. *20*, 157–171.

Ferreira, L., Sánchez-Juanes, F., Porras-Guerra, I., García-García, M.I., García-Sánchez, J.E., González-Buitrago, J.M., and Muñoz-Bellido, J.L. (2011). Microorganisms direct identification from blood culture by matrix-assisted laser desorption/ionization time-of-flight mass spectrometry. Clin. Microbiol. Infect. *17*, 546–551.

Ferroni, A., Suarez, S., Beretti, J.L., Dauphin, B., Bille, E., Meyer, J., Bougnoux, M.E., Alanio, A., Berche, P., and Nassif, X. (2010). Real-time identification of bacteria and *Candida* species in positive blood culture broths by matrix-assisted laser desorption ionization-time of flight mass spectrometry. J. Clin. Microbiol. *48*, 1542–1548.

Firacative, C., Trilles, L., and Meyer, W. (2012). MALDI-TOF MS enables the rapid identification of the major molecular types within the *Cryptococcus neoformans/C. gattii* species complex. PloS One *7*, e37566.

Gabaldón, T., Martin, T., Marcet-Houben, M., Durrens, P., Bolotin-Fukuhara, M., Lespinet, O., Arnaise, S., Boisnard, S., Aguileta, G., Atanasova, R., *et al.* (2013). Comparative genomics of emerging pathogens in the *Candida glabrata* clade. BMC Genomics. *14*, 623.

Gaitanis, G., Magiatis, P., Hantschke, M., Bassukas, I.D., and Velegraki, A. (2012). The *Malassezia* genus in skin and systemic diseases. Clin. Microbiol. Rev. *25*, 106–141.

Gautier, M., Ranque, S., Normand, A.C., Becker, P., Packeu, A., Cassagne, C., L'Ollivier, C., Hendrickx, M., and Piarroux, R. (2014). MALDI-TOF mass spectrometry: revolutionising clinical laboratory diagnosis of mould infections. Clin. Microbiol. Infect. *20*, 1366-1371.

Gorton, R.L., Seaton, S., Ramnarain, P., McHugh, T.D., and Kibbler, C.C. (2014). Evaluation of a short, on-plate formic acid (FA) extraction method for MALDI-TOF MS-based identification of clinically relevant yeast isolates. J. Clin. Microbiol. *52*, 1253–1255.

Gupta, A.K., Boekhout, T., Theelen, B., Summerbell, R., and Batra, R. (2004). Identification and typing of *Malassezia* species by amplified fragment length polymorphism and sequence analyses of the internal transcribed spacer and large-subunit regions of ribosomal DNA. J. Clin. Microbiol. *42*, 4253–4260.

Hagen, F., Khayhan, K., Theelen, B., Kolecka, A., Polacheck, I., Sionov, E., Falk, R., Parnmen, S., Lumbsch, H.T., and Boekhout, T. (2015). Recognition of seven species in the *Cryptococcus gattii/Cryptococcus neoformans* species complex. Fungal Genet. Biol. *78*, 16-48.

Hendrickx, M., Goffinet, J.S., Swinne, D., and Detandt, M., (2011). Screening of strains of the *Candida parapsilosis* group of the BCCM/IHEM collection by MALDI-TOF MS. Diagn. Microbiol. Infect. Dis. *70*, 544–548.

Hettick, J.M., Green, B.J., Buskirk, A.D., Kashon, M.L., Slaven, J.E., Janotka, E., Blachere, F.M., Schmechel, D., and Beezhold, D.H. (2008a). Discrimination of *Penicillium* isolates by matrix-assisted laser desorption/ionization time-of-flight mass spectrometry fingerprinting. Rapid Commun. Mass Spectrom. *22*, 2555–2560.

Hettick, J.M., Green, B.J., Buskirk, A.D., Kashon, M.L., Slaven, J.E., Janotka, E., Blachere, F.M., Schmechel, D., and Beezhold, D.H. (2008b). Discrimination of *Aspergillus* isolates at the species and strain level by matrix-assisted laser desorption/ionization time-of-flight mass spectrometry fingerprinting. Anal. Biochem. *380*, 276281.

Hillenkamp, F., and Karas, M. (1990). Mass spectrometry of peptides and proteins by matrix-assisted ultraviolet laser desorption/ionization. Methods Enzymol. *193*, 80–95.

Holland, R.D., Wilkes, J.G., Rafii, F., Sutherland, J.B., Persons, C.C., Voorhees, K.J., and Lay, J.O. Jr. (1996). Rapid identification of intact whole bacteria based on spectral patterns using matrix-assisted laser desorption/ionization with time-of-flight mass spectrometry. Rapid Commun. Mass Spectrom. *10*, 1227–1232.

Iatta, R., Cafarchia, C., Cuna, T., Montagna, O., Laforgia, N., Gentile, O., Rizzo, A., Boekhout, T., Otranto, D., and Montagna, M.T. (2014). Bloodstream infections by *Malassezia* and *Candida* species in critical care patients. Med. Mycol. *52*, 264–269.

Iriart, X., Lavergne, R.A., Fillaux, J., Valentin, A., Magnaval, J.F., Berry, A., and Cassaing, S. (2012). Routine identification of medical fungi by the new Vitek MS matrix-assisted laser desorption ionization-time of flight system with a new time-effective strategy. J. Clin. Microbiol. *50*, 2107–2110.

Jamal, W.Y., Ahmad, S., Khan, Z.U., and Rotimi, V.O. (2014). Comparative evaluation of two matrix-assisted laser desorption/ionization time-of-flight mass spectrometry (MALDI-TOF MS) systems for the identification of clinically significant yeasts. Int. J. Infect. Dis. *26*, 167–170.

Karas, M., Bachmann, D., Bahr, U., and Hillenkamp, F. (1987). Matrix-assisted ultraviolet laser desorption of non-volatile compounds. F. Int. J. Mass Spectrom. Ion Processes *78*, 53–68.

Kessler, A.T., Kourtis, A.P., and Simon, N. (2002). Peripheral thromboembolism associated with *Malassezia* furfur sepsis. Ped. Infect. Dis. J. *21*, 356–357.

Kolecka, A., Khayhan, K., Arabatzis, M., Velegraki, A., Kostrzewa, M., Andersson, A., Scheynius, A., Cafarchia, C., Iatta, R., Montagna, M.T., *et al.* (2014). Efficient identification of *Malassezia* yeasts by matrix-assisted laser desorption ionization-time of flight mass spectrometry (MALDI-TOF MS). Br. J. Dermatol. *170*, 332–341.

Kolecka, A., Khayhan, K., Groenewald, M., Theelen, B., Arabatzis, M., Velegraki, A., Kostrzewa, M., Mares, M., Taj-Aldeen, S.J., and Boekhout, T. (2013). Identification of medically relevant species of arthroconidial yeasts by use of matrix-assisted laser desorption ionization-time of flight mass spectrometry. J. Clin. Microbiol. *51*, 2491–2500.

Kondori, N., Erhard, M., Welinder-Olsson, C., Groenewald, M., Verkley, G., and Moore, E.R. (2015). Analyses of black fungi by matrix-assisted laser desorption/ionization time-of-flight mass spectrometry (MALDI-TOF MS): species-level identification of clinical isolates of *Exophiala dermatitidis*. FEMS Microbiol. Lett. *362*, 1–6.

Krause, E., Wichels, A., Erler, R., and Gerdts, G. (2013). Study on the effects of near-future ocean acidification on marine yeasts: a microcosm approach. Helgoland Marine Research, *67*, 607–621.

Krishnamurthy, T., Rajamani, U., and Ross, P.L. (1996). Detection of pathogenic and non-pathogenic bacteria by Matrix-assisted Laser Desorption/Ionization Time-of-flight Mass Spectrometry. Rapid Commun. Mass Spectrom. *10*, 883–888.

Kubesová, A., Šalplachta, J., Horká, M., Růžička, F., and Šlais, K. (2012). *Candida 'psilosis'*-electromigration techniques and MALDI-TOF mass spectrometry for phenotypical discrimination. Analyst *137*, 1937–1943.

Kurtzman, C.P., Fell, J.W., Boekhout, T., and Robert, V. (2011). Methods for isolation, phenotypic characterization and maintenance of yeasts. In The Yeasts, a Taxonomic Study, 5th edn, Kurtzman, C.P., Fell, J.W., and Boekhout, T., eds. (Elsevier, Amsterdam, Netherlands), pp. 87–110.

Kurtzman, C.P., and Robnett, C.J. (1998). Identification and phylogeny of ascomycetous yeasts from analysis of nuclear large subunit (26S) ribosomal DNA partial sequences. Antonie Van Leeuwenhoek *73*, 331–371.

Lachance, M.-A., Boekhout, T., Scorzetti, G., Fell, J.W., and Kurtzman, C.P. (2011). Candida Berkhout (1923). In The Yeasts, a Taxonomic Study, 5th edn, Kurtzman, C.P., Fell, J.W., and Boekhout, T., eds. (Elsevier, Amsterdam, Netherlands), pp. 987–1278.

Lacroix, C., Gicquel, A., Sendid, B., Meyer, J., Accoceberry, I., Francois, N., Morio, F., Desoubeaux, G., Chandenier, J., Kauffmann-Lacroix, C., *et al.* (2014). Evaluation of two matrix-assisted laser desorption ionization-time of flight mass spectrometry (MALDI-TOF MS) systems for the identification of *Candida* species. Clin. Microbiol. Infect. *20*, 153–158.

Lau, A.F., Drake, S.K., Calhoun, L.B., Henderson, C.M., and Zelazny, A.M. (2013). Development of a clinically comprehensive database and a simple procedure for identification of molds from solid media by matrix-assisted laser desorption ionization-time of flight mass spectrometry. J. Clin. Microbiol. *51*, 828–834.

Lohmann, C., Sabou, M., Moussaoui, W., Prévost, G., Delarbre, J.M., Candolfi, E., Gravet, A., and Letscher-Bru, V. (2013). Comparison between the Biflex III-Biotyper and the Axima-SARAMIS systems for yeast identification by matrix-assisted laser desorption ionization-time of flight mass spectrometry. J. Clin. Microbiol. *51*, 1231–1236.

Mancini, N., De Carolis, E., Infurnari, L., Vella, A., Clementi, N., Vaccaro, L., Ruggeri, A., Posteraro, B., Burioni, R., Clementi, M., and Sanguinetti, M. (2013). Comparative evaluation of the Bruker Biotyper and Vitek MS matrix-assisted laser desorption ionization-time of flight (MALDI-TOF) mass spectrometry systems for identification of yeasts of medical importance. J. Clin. Microbiol. *51*, 2453–2457.

Marinach, C., Alanio, A., Palous, M., Kwasek, S., Fekkar, A., Brossas, J.Y., Brun, S., Snounou, G., Hennequin, C., Sanglard, D., *et al.* (2009). MALDI-TOF MS-based drug susceptibility testing of pathogens: the example of *Candida albicans* and fluconazole. Proteomics *9*, 4627–4631.

Marinach-Patrice, C., Fekkar, A., Atanasova, R., Gomes, J., Djamdjian, L., Brossas, J.Y., Meyer, I., Buffet, P., Snounou, G., Datry, A., *et al.* (2010). Rapid species diagnosis for invasive candidiasis using mass spectrometry. PLoS One *25*, e8862.

Marklein, G., Josten, M., Klanke, U., Muller, E., Horre, R., Maier, T., Wenzel, T., Kostrzewa, M., Bierbaum, G., Hoerauf, A., *et al.* (2009). Matrix-assisted laser desorption ionization-time of flight mass spectrometry for fast and reliable identification of clinical yeast isolates. J. Clin. Microbiol. *47*, 2912–2917.

McCullough, M.J., Clemons, K.V., Farina, C., McCusker, J.H., and Stevens, D.A. (1998). Epidemiological investigation of vaginal *Saccharomyces cerevisiae* isolates by a genotypic method. J. Clin. Microbiol. *36*, 557–62. Erratum in: J. Clin. Microbiol. 2000, *38*, 1311.

McTaggart, L.R., Lei, E., Richardson, S.E., Hoang, L., Fothergill, A., and Zhang, S.X. (2011). Rapid identification of *Cryptococcus neoformans* and *Cryptococcus gattii* by matrix-assisted laser desorption ionization-time of flight mass spectrometry. J. Clin. Microbiol. *49*, 3050–3053.

Minea, B., Nastasa, V., Moraru, R.F., Kolecka, A., Flonta, M.M., Marincu, I., Man, A., Toma, F., Lupse, M., Doroftei, B., *et al.* (2015). Species distribution and susceptibility profile to fluconazole, voriconazole and MXP-4509 of 551 clinical yeast isolates from a Romanian multi-centre study. Eur. J. Clin. Microbiol. Infect. Dis. *34*, 367–383.

Moothoo-Padayachie, A., Kandappa, H.R., Krishna, S.B.N., Maier, T., and Govender, P. (2012). Biotyping *Saccharomyces cerevisiae* strains using matrix-assisted laser desorption/ionization time-of-flight mass spectrometry (MALDI-TOF MS). Eur. Food. Res. Technol. *236*, 351–364.

Muñoz, P., Bouza, E., Cuenca-Estrella, M., Eiros, J.M., Pérez, M.J., Sánchez-Somolinos, M., Rincón, C., Hortal, J., and Peláez, T. (2005). *Saccharomyces cerevisiae* fungemia: an emerging infectious disease. Clin. Infect. Dis. *40*, 1625–1634.

Nenoff, P., Erhard, M., Simon, J.C., Muylowa, G.K., Herrmann, J., Rataj, W., and Gräser, Y. (2013). MALDI-TOF mass spectrometry – a rapid method for the identification of dermatophyte species. Med. Mycol. *51*, 17–24.

Nguyen, S.T., Lund, C.H., and Durand, D.J. (2001). Thrombolytic therapy for adhesion of percutaneous central venous catheters to vein intima associated with *Malassezia furfur* Infection. *21*, 331–333.

Nilsson, R.H., Ryberg, M., Kristiansson, E., Abarenkov, K., Larsson, K.-H., and Kõljalg, U. (2006). Taxonomic reliability of dna sequences in public sequence databases: a fungal perspective. PLoS One *1*, e59.

Normand, A.C., Cassagne, C., Ranque, S., L'Ollivier, C., Fourquet, P., Roesems, S., Hendrickx, M., and Piarroux, R. (2013). Assessment of various parameters to improve MALDI-TOF MS reference spectra libraries constructed for the routine identification of filamentous fungi. BMC Microbiol. *13*, 76.

Oliveri, S., Trovato, L., Betta, P., Romeo, M.G., and Nicoletti, G. (2011). *Malassezia furfur* fungaemia in a neonatal patient detected by lysis-centrifugation blood culture method: first case reported in Italy. Mycoses. *54*, e638–40.

Packeu, A., Hendrickx, M., Beguin, H., Martiny, D., Vandenberg, O., and Detandt, M. (2013). Identification of the *Trichophyton mentagrophytes* complex species using MALDI-TOF mass spectrometry. Med. Mycol. *51*, 580–585.

Patel, R. (2015). MALDI-TOF MS for the diagnosis of infectious diseases. Clin. Chem. *61*, 100–111.

Pavlovic, M., Mewes, A., Maggipinto, M., Schmidt, W., Messelhäußer, U., Balsliemke, J., Hörmansdorfer, S., Busch, U., and Huber, I. (2014). MALDI-TOF MS based identification of food-borne yeast isolates. J. Microbiol. Methods *106*, 123–128.

Pfaller, M.A., Andes, D.R., Diekema, D.J., Horn, D.L., Reboli, A.C., Rotstein, C., Franks, B., and Azie, N.E. (2014). Epidemiology and outcomes of invasive candidiasis due to non-*albicans* species of *Candida* in 2,496 patients: Data from the Prospective Antifungal Therapy (PATH) Registry 2004–2008. PLoS One *9*, e101510.

Pinto, A., Halliday, C., Zahra, M., van Hal, S., Olma, T., Maszewska, K., Iredell, J.R., Meyer, W., and Chen, S.C. (2011). Matrix-assisted laser desorption ionization-time of flight mass spectrometry identification of yeasts is contingent on robust reference spectra. PloS One *6*, e25712.

Posteraro, B., De Carolis, E., Vella, A., and Sanguinetti, M. (2013). MALDI-TOF mass spectrometry in the clinical mycology laboratory: identification of fungi and beyond. Expert Rev. Proteomics *10*, 151–164.

Posteraro, B., Torelli, R., De Carolis, E., Posteraro, P., and Sanguinetti, M. (2014). Antifungal susceptibility testing: current role from the clinical laboratory perspective. Mediterr. J. Hematol. Infect. Dis. *6*, e2014030.

Posteraro, B., Vella, A., Cogliati, M., De Carolis, E., Florio, A.R., Posteraro, P., Sanguinetti, M., and Tortorano, A.M. (2012). Matrix-assisted laser desorption ionization-time of flight mass spectrometry-based method for discrimination between molecular types of *Cryptococcus neoformans* and *Cryptococcus gattii*. J. Clin. Microbiol. *50*, 2472–2476.

Pulcrano, G., Roscetto, E., Iula, V.D., Panellis, D., Rossano, F., and Catania, M.R. (2012). MALDI-TOF mass spectrometry and microsatellite markers to evaluate *Candida parapsilosis* transmission in neonatal intensive care units. Eur. J. Clin. Microbiol. Infect. Dis. *31*, 2919–2928.

Quiles-Melero, I., Garcia-Rodriguez, J., Gomez-Lopez, A., and Mingorance, J. (2012). Evaluation of matrix-assisted laser desorption/ionisation time-of-flight (MALDI-TOF) mass spectrometry for identification of *Candida parapsilosis, C. orthopsilosis* and *C. metapsilosis*. Eur. J. Clin. Microbiol. Infect. Dis. *31*, 67–71.

Ramos, L.S., Figueiredo-Carvalho, M.H., Barbedo, L.S., Ziccardi, M., Chaves, A.L., Zancopé-Oliveira, R.M., Pinto, M.R., Sgarbi, D.B., Dornelas-Ribeiro, M., Branquinha, M.H., *et al.* (2015). *Candida haemulonii* complex: species identification and antifungal susceptibility profiles of clinical isolates from Brazil. J. Antimicrob. Chemother. *70*, 111–115.

Revankar, S.G., and Sutton, D.A. (2010). Melanized fungi in human disease. Clin. Microbiol. Rev. *23*, 884–928.

Riat, A., Rentenaar, R.J., van Drongelen, A.M., Barras, V., Bertens, L.C., Vlek, A.L., Doppenberg, E., Weersink, A.J., Reinders, E., Vlaminckx, B.J., *et al.* (2015). Ground steel target plates in combination with direct transfer of clinical yeast isolates improves frequencies of species level MALDI-TOF MS identifications in comparison with polished steel target plates. J. Clin. Microbiol. 53, 1993–1995.

Riquelme, A.J., Calvo, M.A., Guzmán, A.M., Depix, M.S., García, P., Pérez, C., Arrese, M., and Labarca, J.A. (2003). *Saccharomyces cerevisiae* fungemia after *Saccharomyces boulardii* treatment in immunocompromised patients. J. Clin. Gastroenterol. *36*, 41–43.

Romanelli, A.M., Sutton, D.A., Thompson, E.H., Rinaldi, M.G., and Wickes, B.L. (2010). Sequence-based identification of filamentous basidiomycetous fungi from clinical specimens: a cautionary note. J. Clin. Microbiol. *48*, 741–752.

Romeo, O., and Criseo, G. (2008). First molecular method for discriminating between *Candida africana, Candida albicans*, and *Candida dubliniensis* by using HWP1 gene. Diagn. Microbiol. Infect. Dis. *62*, 230–233.

Romeo, O., and Criseo, G. (2009). Morphological, biochemical and molecular characterisation of the first Italian *Candida africana* isolate. Mycoses 52, 454–457.

Rosenvinge, F.S., Dzajic, E., Knudsen, E., Malig, S., Andersen, L.B., Løvig, A., Arendrup, M.C., Jensen, T.G., Gahrn-Hansen, B., and Kemp, M. (2013). Performance of matrix-assisted laser desorption-time of flight mass spectrometry for identification of clinical yeast isolates. Mycoses *56*, 229–235.

Samson, R.A., Houbraken, J., Thrane, U., Frisvad, J.C., and Andersen, B. (2010). CBS Laboratory Manual Series 2: Food and indoor fungi (Utrecht, the Netherlands: CBS-KNAW Fungal Biodiversity Centre).

Sanguinetti, M., and Posteraro, B. (2014). MALDI-TOF mass spectrometry: any use for *Aspergilli*? Mycopathologia *178*, 417–426.

Santos, C., Lima, N., Sampaio, P., and Pais, C., (2011). Matrix-assisted laser desorption/ionization time-of-flight intact cell mass spectrometry to detect emerging pathogenic *Candida* species. Diagn. Microbiol. Infect. Dis. *71*, 304–308.

Santos, C., Paterson, R.R., Venâncio, A., and Lima, N. (2010). Filamentous fungal characterizations by matrix-assisted laser desorption/ionization time-of-flight mass spectrometry. J. Appl. Microbiol. *108*, 375–385.

Schieffer, K.M., Tan, K.E., Stamper, P.D., Somogyi, A., Andrea, S.B., Wakefield, T., Romagnoli, M., Chapin, K.C., Wolk, D.M., and Carroll, K.C. (2014). Multicenter evaluation of the Sepsityper™ extraction kit and MALDI-TOF MS for direct identification of positive blood culture isolates using the BD BACTEC™ FX and VersaTREK(®) diagnostic blood culture systems. J. Appl. Microbiol. *116*, 934–941.

Schleman, K.A., Tullis, G., and Blum, R. (2000). Intracardiac mass complicating *Malassezia furfur* fungemia. Chest. *118*, 1828–1829.

Schoch, C.L., Seifert, K.A., Huhndorf, S., Robert, V., Spouge, J.L., Levesque, C.A., and Chen, W. (2012). Nuclear ribosomal internal transcribed spacer (ITS) region as a universal DNA barcode marker for Fungi. Proc. Natl. Acad. Sci. USA *109*, 6241–6246.

Schulthess, B., Ledermann, R., Mouttet, F., Zbinden, A., Bloemberg, G.V., Böttger, E.C., and Hombach, M. (2014). Use of the Bruker MALDI Biotyper for identification of molds in the clinical mycology laboratory. J. Clin. Microbiol. *52*, 2797–2803.

Sendid, B., Ducoroy, P., Francois, N., Lucchi, G., Spinali, S., Vagner, O., *et al.* (2013). Evaluation of MALDI-TOF mass spectrometry for the identification of medically-important yeasts in the clinical laboratories of Dijon and Lille hospitals. Med. Mycol. *51*, 25–32.

Setlhare, G., Malebo, N., Shale, K., and Lues, R. (2014). Identification of airborne microbiota in selected areas in a health-care setting in South Africa. BMC Microbiol. *22*, 100.

Seyedmousavi, S., Guillot, J., and de Hoog, G.S. (2013). Phaeohyphomycoses, emerging opportunistic diseases in animals. Clin. Microbiol. Rev. *26*, 19–35.

Seyedmousavi, S., Netea, M.G., Mouton, J.W., Melchers, W.J., Verweij, P.E., and de Hoog, G.S. (2014). Black yeasts and their filamentous relatives: principles of pathogenesis and host defense. Clin. Microbiol. Rev. *27*, 527–542.

Seyfarth, F., Wiegand, C., Erhard, M., Graser, Y., Elsner, P., and Hipler, U.C. (2012). Identification of yeast isolated from dermatological patients by MALDI-TOF mass spectrometry. Mycoses *55*, 276–280.

Spanu, T., Posteraro, B., Fiori, B., D'Inzeo, T., Campoli, S., Ruggeri, A., Tumbarello, M., Canu, G., Trecarichi, E.M., Parisi, G., Tronci, M., Sanguinetti, M., and Fadda, G. (2012). Direct maldi-tof mass spectrometry assay of blood culture broths for rapid identification of *Candida* species causing bloodstream infections: an observational study in two large microbiology laboratories. J. Clin. Microbiol. *50*, 176–179.

Spinali, S., van Belkum, A., Goering, R.V., Girard, V., Welker, M., Van Nuenen, M., Pincus, D.H., Arsac, M., and Durand, G. (2015). Microbial typing by Matrix-Assisted Laser Desorption Ionization-Time of Flight Mass Spectrometry: do we need guidance for data interpretation? J. Clin. Microbiol. *53*, 760–765.

Stevenson, L.G., Drake, S.K., Shea, Y.R., Zelazny, A.M., and Murray, P.R. (2010). Evaluation of matrix-assisted laser desorption ionization-time of flight mass spectrometry for identification of clinically important yeast species. J. Clin. Microbiol. *48*, 3482–3486.

Sugita, T., Takashima, M., Poonwan, N., and Mekha, N. (2006). *Candida pseudohaemulonii* Sp. Nov., an amphotericin B-and azole-resistant yeast species, isolated from the blood of a patient from Thailand. Microbiol. Immunol. *50*, 469–473.

Sullivan, D.J., Westerneng, T.J., Haynes, K.A., Bennett, D.E., and Coleman, D.C. (1995). *Candida dubliniensis* sp. nov.: phenotypic and molecular characterization of a novel species associated with oral candidosis in HIV-infected individuals. Microbiology. *141*, 1507–1521.

Taj-Aldeen, S.J., Al-Ansari, N., El Shafei, S., Meis, J.F., Curfs-Breuker, I., Theelen, B., and Boekhout, T. (2009). Molecular identification and susceptibility of *Trichosporon* species isolated from clinical specimens in Qatar: isolation of *Trichosporon dohaense* Taj-Aldeen, Meis & Boekhout sp. nov. J. Clin. Microbiol. *47*, 1791–1799.

Taj-Aldeen, S.J., Kolecka, A., Boesten, R., Alolaqi, A., Almaslamani, M., Chandra, P., Meis, J.F., and Boekhout, T. (2013). Epidemiology of candidemia in Qatar, the Middle East: performance of MALDI-TOF MS for the identification of *Candida* species, species distribution, outcome, and susceptibility pattern. Infection *42*, 393–404.

Tan, K.E., Ellis, B.C., Lee, R., Stamper, P.D., Zhang, S.X., and Carroll, K.C. (2012). Prospective evaluation of a matrix-assisted laser desorption ionization-time of flight mass spectrometry system in a hospital clinical microbiology laboratory for identification of bacteria and yeasts: a bench-by-bench study for assessing the impact on time to identification and cost-effectiveness. J. Clin. Microbiol. *50*, 3301–3308.

Tanaka, K., Waki, H., Ido, Y., Akita, S., Yoshida, Y., Yoshida, T., and Matsuo, T. (1988). Protein and polymer analyses up to m/z 100 000 by laser ionization time-of-flight mass spectrometry. Rapid Commun. Mass Spectrom. *2*, 151–153.

Theel, E.S., Hall, L., Mandrekar, J., and Wengenack, N.L. (2011). Dermatophyte identification using matrix-assisted laser desorption ionization-time of flight mass spectrometry. J. Clin Microbiol. *49*, 4067–4071.

Tietz, H.J., Hopp, M., Schmalreck, A., Sterry, W., and Czaika, V. (2001). *Candida africana* sp. nov., a new human pathogen or a variant of *Candida albicans*? Mycoses. *44*, 437–445.

Tietz, H.J., Küssner, A., Thanos, M., De Andrade, M.P., Presber, W., and Schönian, G. (1995). Phenotypic and genotypic characterization of unusual vaginal isolates of *Candida albicans* from Africa. J. Clin. Microbiol. *33*, 2462–2465.

Usbeck, J.C., Kern, C.C., Vogel, R.F., and Behr, J. (2013). Optimization of experimental and modelling parameters for the differentiation of beverage spoiling yeasts by Matrix-Assisted-Laser-Desorption/Ionization-Time-of-Flight Mass Spectrometry (MALDI-TOF MS) in response to varying growth conditions. Food Microbiol. *36*, 379–387.

Usbeck, J.C., Wilde, C., Bertrand, D., Behr, J., and Vogel, R.F. (2014). Wine yeast typing by MALDI-TOF MS. Appl. Microbiol. Biotechnol. *98*, 3737–3752.

Vallejo, J.A., Miranda, P., Flores-Félix, J.D., Sánchez-Juanes, F., Ageitos, J.M., González-Buitrago, J.M., Velázquez, E., and Villa, T.G. (2013). Atypical yeasts identified as *Saccharomyces cerevisiae* by MALDI-TOF MS and gene sequencing are the main responsible of fermentation of chicha, a traditional beverage from Peru. Syst. Appl. Microbiol. *36*, 560–564.

van Herendael, B.H., Bruynseels, P., Bensaid, M., Boekhout, T., De Baere, T., *et al.* (2012). Validation of a modified algorithm for the identification of yeast isolates using matrix-assisted laser desorption/ionisation time-of-flight mass spectrometry (MALDI-TOF MS). Eur. J. Clin. Microbiol. Infect. Dis. *31*, 841–848.

van Veen, S.Q., Claas, E.C., and Kuijper, E.J. (2010). High-throughput identification of bacteria and yeast by matrix-assisted laser desorption ionization-time of flight mass spectrometry in conventional medical microbiology laboratories. J. Clin. Microbiol. *48*, 900–907.

Vaughan-Martini A., and Martini, A. (2011). Saccharomyces Meyen ex Reess (1870). In The Yeasts, a Taxonomic Study, 5th edn, Kurtzman, C.P., Fell, J.W., and Boekhout, T., eds. (Elsevier, Amsterdam, Netherlands), pp. 733–746.

Vella, A., De Carolis, E., Vaccaro, L., Posteraro, P., Perlin, D.S., Kostrzewa, M., Posteraro, B., and Sanguinetti, M. (2013). Rapid antifungal susceptibility testing by matrix-assisted laser desorption ionization-time of flight mass spectrometry analysis. J. Clin. Microbiol. *51*, 2964–2969.

Vermeulen, E., Verhaegen, J., Indevuyst, C., and Lagrou, K. (2012). Update on the evolving role of MALDI-TOF MS for laboratory diagnosis of fungal infections. Curr. Fungal Infect. Rep. *6*, 206–214.

Videira, S.I.R., Groenewald, J.Z., Kolecka, A., van Haren, L., Boekhout, T., and Crous, P.W. (2015). Elucidating the *Ramularia eucalypti* species complex. Persoonia *34*, 50–64.

Vilgalys, R. (2003). Taxonomic misidentification in public DNA databases. New Phytologist. *160*, 4–5.

Vlek, A., Kolecka, A., Khayhan, K., Theelen, B., Groenewald, M., Boel, E., Multicenter Study Group, and Boekhout, T. (2014). Interlaboratory comparison of sample preparation methods, database expansions, and cutoff values for identification of yeasts by matrix-assisted laser desorption ionization-time of flight mass spectrometry using a yeast test panel. J. Clin. Microbiol. *52*, 3023–3029.

Welham, K.J., Domin, M.A., Johnson, K., Jones, L., and Ashton, D.S. (2000). Characterization of fungal spores by laser desorption/ionization time-of-flight mass spectrometry. Rapid Commun. Mass Spectrom. *14*, 307–310.

Westblade, L.F., Jennemann, R., Branda, J.A., Bythrow, M., Ferraro, M.J., Garner, O.B., Ginocchio, C.C., Lewinski, M.A., Manji, R., Mochon, A.B., *et al.* (2013). Multi-center study evaluating the VITEK MS for identification of medically important yeasts. J. Clin. Microbiol. *51*, 2267–2272.

White, T.J., Lee, B.T.S., and Taylor, J. (1985). Amplification and direct sequencing of fungal ribosomal RNA genes for phylogenetics. PCR Protocols 3, 315–322.

Willinger, B., and Haase, G. (2013). State-of-the-art procedures and quality management in diagnostic medical mycology. Curr. Fungal Infect. Rep. 7, 260–272.

Yamamoto, M., Umeda, Y., Yo, A., Yamaura, M., and Makimura, K. (2014). Utilization of matrix-assisted laser desorption and ionization time-of-flight mass spectrometry for identification of infantile seborrheic dermatitis-causing *Malassezia* and incidence of culture-based cutaneous *Malassezia* microbiota of 1-month-old infants. J. Dermatol. *41*, 117–123.

Yan, Y., He, Y., Maier, T., Quinn, C., Shi, G., Li, H., Stratton, C.W., Kostrzewa, M., and Tang, Y.W. (2011). Improved identification of yeast species directly from positive blood culture media by combining Sepsityper specimen processing and Microflex analysis with the matrix-assisted laser desorption ionization Biotyper system. J. Clin. Microbiol. *49*, 2528–2532.

Molecular Typing of Bacteria/ Fungi Using MALDI-TOF MS

4

Silpak Biswas, Frédérique Gouriet and Jean-Marc Rolain

Abstract

Traditional microbial typing technologies for the characterization of pathogenic microorganisms and the monitoring of their global spread are often difficult to standardize and lack sufficient ease of use. Timely reporting of strain typing information is essential for the early initiation of infection control measures to prevent a further dissemination of the pathogen and for epidemiological purposes. Matrix-assisted laser desorption ionization time-of-flight mass spectrometry (MALDI-TOF MS) has become an important method in clinical microbiology laboratories for identifying and typing bacteria and fungi. With the exception of a few difficult strains, MALDI-TOF MS using standardized procedures allows the accurate identification of Gram-positive and Gram-negative bacterial strains at the species and subspecies levels. Recent studies have shown that MALDI-TOF MS also has the potential to accurately identify filamentous fungi. MALDI-TOF MS is a powerful tool for species and subspecies classification and rapid identification and typing of pathogenic microbes. Furthermore, the implementation of specific databases of well-known pathogens will likely increase the role of MALDI-TOF MS for typing in the future.

Introduction

Accurate bacterial identification and typing are important in the case of outbreaks of infectious diseases and play crucial roles in diagnosis and efficient treatment. Matrix-assisted laser desorption/ionization time-of-flight mass spectrometry (MALDI-TOF MS) has become an important technology for the identification of bacteria in pharmaceutical and medical microbiology laboratories (Seng *et al.*, 2009, 2010; Bizzini and Greub, 2010; Croxatto *et al.*, 2012). Because MALDI-TOF MS detects a large spectrum of proteins, the technique is able to discriminate between closely related species and to classify organisms at the species level (Seng *et al.*, 2010). Moreover, MALDI-TOF MS has been used successfully for microbial typing and identification at the subspecies level (Dieckmann *et al.*, 2008; Williamson *et al.*, 2008; Lartigue *et al.*, 2009; Cherkaoui *et al.*, 2010). In addition, the capability of MALDI-TOF MS to rapidly characterize microorganisms favours its potential applications in multiple areas, including medical diagnostics, bio-defence, environmental monitoring, and food quality control. Several studies have shown that the requirements for MALDI-TOF MS-dependent microbial typing are different and more complex than those required for routine microbial identification. Indeed, microbial typing at the subspecies level requires very different sample preparation and analytical procedures (Emonet *et al.*, 2010; Murray, 2010), and the rigorous optimization of testing parameters is crucial for strain typing.

For epidemiological studies, the typing of several microorganisms, such as *Staphylococcus* and *Listeria* species, requires the use of conventional techniques such as pulsed-field gel electrophoresis (PFGE), amplified fragment length polymorphism analysis (AFLP) and multilocus sequence analysis (MLSA) (Enright *et al.*, 2000; Jackson *et al.*, 2005; Cherkaoui *et al.*, 2010; Harmsen *et al.*, 2003). Conventional identification approaches, such as Gram staining, culture and growth characteristics, biochemical tests, multiple susceptibility testing, and serotyping, are important methods for understanding the epidemiology of community- and healthcare-associated infections, though these tests have some specific limitations and require longer time frames for bacterial identification. Such time-consuming procedures affect the proper treatment of patients with respect to antibiotic and supportive treatments. Hence, for the past several years, conventional phenotypic identification has been replaced by the genetic typing of species in medical microbiology laboratories (Bizzini and Greub, 2010; Seng *et al.*, 2010; Cherkaoui *et al.*, 2010). Only a few hours are required to obtain molecular typing results by MALDI-TOF MS, whereas PFGE data could be obtained after several days. Reproducibility, speed, and sensitivity of analyses are major advantages of the MALDI-TOF MS method.

Pathogen identification is crucial to confirm bacterial infections and to guide antimicrobial therapy. MALDI-TOF-MS is a rapid and inexpensive technology that has the potential to replace or complement conventional phenotypic identification for most bacterial and fungal strains isolated in clinical microbiology laboratories (Emonet *et al.*, 2010; Sogawa *et al.*, 2011, 2012). Currently, MALDI-TOF MS can be used for the accurate and rapid identification of various microorganisms, such as Gram-positive bacteria, *Enterobacteriaceae*, non-fermenting bacteria, mycobacteria, anaerobes, and fungi (van Veen *et al.*, 2010; Biswas and Rolain, 2013; Barbuddhe *et al.*, 2008; Degand *et al.*, 2008; Mellmann *et al.*, 2008; Pignone *et al.*, 2006; Bittar *et al.*, 2010). In a short time, the technique has been widely adopted and integrated into many medical microbiology laboratories. By testing colonies, only a few minutes are required for a correct identification, enabling the identification of microorganisms at the species levels, also at the sub-species and strains levels and thereby allowing the detection of epidemic lineages (Murray, 2010).

MALDI-TOF MS has already been used to identify Gram-positive rods and cocci, *Enterobacteriaceae* and Gram-negative rods with acceptable results (Seng *et al.*, 2009, 2010; Croxatto *et al.*, 2012; Spinali *et al.*, 2015). The rapid and accurate identification of type species of bacterial isolates is essential for epidemiological surveillance and infection control studies. The potential advantages that MALDI-TOF-MS offers over other techniques of microbial characterization include minimal sample preparation, rapid results, and negligible reagent costs (Emonet *et al.*, 2010; Seng *et al.*, 2010). Using optimal sample preparation and MALDI conditions for discrimination at the strain level in combination with the Pearson coefficient, Vargha *et al.* (2006) showed that MALDI-TOF MS offers a better discriminatory power than 16S rRNA gene sequencing for the classification of *Arthrobacter* isolates at the subspecies level. Here, we review the current literature on the use of the MALDI-TOF technique for the typing of microorganisms.

MALDI-TOF-MS technique

Each mass spectrometer consists of three functional units: an ion source to ionize and transfer analyte ions into the gas phase; a mass analyser to separate ions by their mass-to-charge

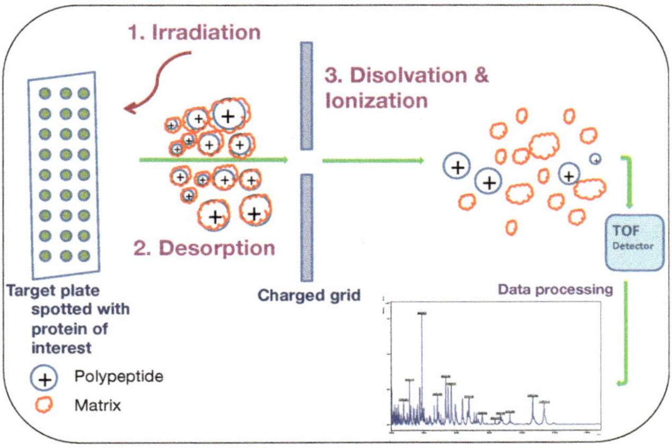

Figure 4.1 MALDI-TOF mass spectrometry.

ratio (m/z); and a detection device to monitor ions (Sauer and Kliem, 2010). Samples are prepared by mixing with a matrix, which results in the crystallization of the sample within the matrix. The composition of the matrix varies according to the biomolecule to be analysed and the type of laser used, and the size and intensities of the peaks of the detected molecules are dependent on the matrix selected for the experiment. Because MALDI mostly generates singly charged ions, the derived spectra may include a larger number of proteins (Emonet *et al.*, 2010; Croxatto *et al.*, 2012) (Fig. 4.1).

The peptide or protein mass-to-charge (m/z) values form the mass spectral peaks, indicating the molecular masses and charge densities of the components present in a biological sample (Sauer and Kliem, 2010). These spectra can generate pathognomonic patterns that provide unbiased identifications of particular species and even genotypes within species. A key requirement for the successful application of MALDI-TOF MS and other proteomic strategies is the assembly of mass-to-charge databases that allow experimental data to be characterized based on profile matching (Cherkaoui *et al.*, 2010). The required performance expected from a mass analyser depends on the type of sample to be analysed and the ultimate goal of the analysis (e.g. quantification, protein identification, microorganism identification, biotyping). Theoretically, because MALDI-TOF MS detects a large spectrum of proteins, the technique should be able to discriminate between closely related species and to classify organisms at the subspecies level. Although only a few biomarkers are useful for species identification, typing would require the analysis of more reproducible peaks and/or comparison to an extended database of spectra that could be implemented at a local hospital for surveillance and typing (Dieckmann *et al.*, 2008; Croxatto *et al.*, 2012).

Use of MALDI-typing for Gram-positive bacteria

MALDI-typing of *Staphylococcus aureus*

An important advantage of MALDI-TOF MS is the rapid identification of *Staphylococcus aureus* and separate species belonging to CoNS (coagulase-negative staphylococci).

Staphylococcus aureus is one of the most frequently isolated nosocomial pathogens and is responsible for a variety of infections, ranging from benign superficial skin infections to life-threatening diseases (Lowy, 1998). MRSA (methicillin-resistant *S. aureus*) infections are associated with increased morbidity and mortality, causing major problems within hospitals worldwide (Cosgrove and Carmeli, 2003). Therefore, effective infection control measures are essential to limit the spread of this pathogen, and MRSA typing is an essential prerequisite for the initiation of targeted hospital infection control measures. Molecular typing methods, such as *S. aureus* protein A (spa) typing, multilocus sequence typing (MLST) and pulsed-field gel electrophoresis (PFGE), have primarily been used for this purpose (Enright *et al.*, 2000; Harmsen *et al.*, 2003). However, these methods are time-consuming and cost intensive. Several studies have described reliable species-level identification of *Staphylococcus* and sub-typing based on MALDI-TOF fingerprints (Edwards-Jones *et al.*, 2000; Walker *et al.*, 2002; Jackson *et al.*, 2005).

Wolters *et al.* used MALDI-TOF MS to investigate the potential discrimination of major MRSA lineages (Wolters *et al.*, 2011), demonstrating that MALDI-TOF MS has the potential to become a valuable first-line tool for the inexpensive and rapid typing of MRSA in infection control. Based on the MALDI-TOF analysis of crude bacterial extracts, these authors reported the establishment and validation of a typing scheme for MRSA covering the most abundant HA-MRSA lineages. They analysed mass spectra from 25 MRSA isolates belonging to the five major hospital-acquired (HA) MRSA clonal complexes (CC5, CC8, CC22, CC30, CC45; deduced from spa typing), and reproducible spectrum differences were observed for 13 characteristic m/z values. Furthermore, the discriminatory indices of MALDI-typing and spa-typing were found to be comparable. Thus, it appears reasonable to assume that MALDI-TOF fingerprinting could significantly improve MRSA surveillance by allowing routine real-time typing. In another study, the peak patterns of 401 MRSA and methicillin-susceptible *S. aureus* (MSSA) strains, including clinical and laboratory strains, were analysed using MALDI-TOF MS (Josten *et al.*, 2013). The results showed that the MALDI-TOF MS signals correspond to (i) more or less conserved housekeeping proteins, e.g. ribosomal proteins, and (ii) other peptides, e.g. stress proteins and low-molecular-weight toxins, and that these peptides sometimes appear in multiple variants that allow for the detection of subgroups of strains. This study supports that MALDI-TOF MS represents a rapid technology for the discrimination of different *S. aureus* clonal lineages. In another study by Josten *et al.*, evaluation of a collection of clinical agr-positive MRSA and MSSA isolates and type strains showed that, using a detection window of m/z 2411–2419, the PSM-mec is detected by mass spectrometry of whole cells with a sensitivity of 0.95 and a specificity of 1, thereby enabling rapid identification of a subgroup of MRSA with a method that is used during routine identification procedures (Josten *et al.*, 2014).

The study of Boggs *et al.* (2012) reported that MALDI-TOF MS can be used to identify *Staphylococcus aureus* strain type USA300, an important human pathogen at the strain level. A genetic algorithm model using ClinProTools software (Bruker Daltonics) was built using 47 isolates of USA300 and 77 non-USA300 *S. aureus* isolates. Three mass/charge peaks (5932, 6423 and 6592) were found to be discriminators between the groups of isolates.

MALDI-typing of *Streptococcus* spp.

Streptococcus pneumoniae (pnc) is the causative agent of many infectious diseases, including pneumonia, septicaemia, otitis media and conjunctivitis. Williamson *et al.* (2008) described a differential proteomic analysis of representative Pnc conjunctival (cPnc) US outbreak isolates using MALDI-TOF MS. The AB 4700 Proteomics Analyser was used to acquire mass spectra, and the analysis of 25 *S. pneumoniae* strains showed that MALDI-TOF MS can differentiate conjunctival *S. pneumoniae* (cPnc) from non-conjunctival controls. Through statistical algorithms and hierarchal clustering, it was demonstrated that the cPnc outbreak isolates from California and the North-eastern United States were very similar. Based on their MALDI-TOF MS fingerprints, putative peptide/protein biomarkers were tentatively identified, one of which was common and exclusively expressed in cPnc isolates. These cPnc proteomic signatures or biomarker candidates could ultimately be fruitful in the diagnosis of this infection.

In another study, MALDI-TOF MS identified a 6250-Da protein specific to sequence type-1 (ST-1) strains and a 7625-Da protein specific to ST-17 strains when used for the identification of group B streptococci (GBS) (Lartigue *et al.*, 2011). These strains are major causes of meningitis and late-onset disease in neonates. The samples were analysed by MALDI-TOF MS using an Ultraflex TOF/TOF III in positive linear mode (Bruker Daltonics, Bremen, Germany). MALDI-TOF MS was found to be an effective tool for typing GBS strains.

Streptococcus pyogenes is one of the most common bacterial pathogens and causes a variety of diseases, ranging from mild and quite frequent non-invasive infections of the upper respiratory tract and skin to severe invasive infections that include necrotizing fasciitis and streptococcal toxic shock syndrome (Facklam, 2002). A mass spectral fingerprinting and proteomic approach using MALDI-TOF MS was applied to detect and identify protein biomarkers of group A *Streptococcus* (GAS) strains, though the study (Moura *et al.*, 2008) used a limited number of GAS isolates. Mass spectra were acquired using a MALDI-TOF/TOF mass spectrometer (AB 4700 Proteomics Analyser) equipped with a nitrogen laser at 337 nm and with a 200 Hz repetition rate. A subset of common, characteristic, and reproducible biomarkers in the range of 2000–14,000 Da were detected. Despite sharing the same emm type, GAS isolates from cases of necrotizing fasciitis were found to be clustered together and distinct from isolates associated with non-invasive infections.

MALDI-typing of *Clostridium difficile*

Clostridium difficile infection (CDI) is a common cause of diarrhoea in hospitalized patients as well as in the community (Mitchell and Gardner, 2012). Owing to the higher toxin production of *C. difficile* ribotype 027, infections result in increased mortality (McDonald *et al.*, 2005). To date, PCR-ribotyping has been the most effective and widely accepted molecular tool for typing *C. difficile* strains. However, MALDI-TOF MS provides an easy-to-handle system for identifying different pathogenic *C. difficile* strains. Reil *et al.* (2011) performed PCR-ribotyping and the typing of different *C. difficile* strains via the extended SARAMIS™ MALDI-TOF system and found specific markers for ribotypes 001, 027 and 126/078, which allowed clonal identification. In that study, MALDI-TOF MS for typing was performed according to the internal standard operating procedure at the laboratory of AnagnosTec (Association for analytical biochemistry and diagnostics GmbH, Zossen, Germany). Using a standard set of 25 different *C. difficile* PCR ribotypes, a database was produced from different mass spectra recorded in the SARAMIS™ software (AnagnosTec, Zossen, Germany). The database was validated with 355 *C. difficile* strains belonging to 29 different PCR ribotypes collected prospectively from all submitted faeces samples in 2009.

The authors of this study identified specific markers for the most frequent *C. difficile* ribotype (001) in Southern Germany, including the highly virulent strain NAP1/027 (Tuller *et al.*, 2011; Reil *et al.*, 2011). Therefore, in the future MALDI-typing will provide a suitable tool for *C. difficile* strains and the surveillance of CDI.

In their recent study by Rizzardi and Akerlund (2015), *Clostridium difficile* strains were typed by MALDI-TOF method, high molecular weight typing, and compared to PCR ribotyping. Among 500 isolates representing 59 PCR ribotypes a total of 35 high molecular weight types could be resolved. Although less discriminatory than PCR ribotyping, the method is extremely fast and simple, and supports for cost-effective screening of isolates during outbreak situations.

MALDI-typing of *Bacillus* spp.

Members of the genus *Bacillus* are rod-shaped bacteria that can be characterized as endospore-forming, obligate or facultative aerobes. *Bacillus* species may be divided into five or six groups (groups I–VI), based on 16S rRNA phylogeny or phenotypic features, respectively (Priest, 1993). Pathogenicity among *Bacillus* spp. is mainly a feature of organisms belonging to the *B. cereus* group, a subgroup of the *B. subtilis* group (group II) within the *Bacillus* genus, e.g. *B. cereus* and *B. anthracis*. *B. anthracis*, is the causative agent of anthrax and is highly relevant to human and animal health. The report by Lasch *et al.* demonstrated the applicability of a combination of MALDI-TOF MS and chemometrics for the rapid and reliable identification of the vegetative cells of the causative agent of anthrax: *Bacillus anthracis* (Lasch *et al.*, 2009). Mass spectra of 102 *B. anthracis* isolates, 121 *B. cereus* isolates and 151 other Bacillus and related genera isolates were collected using an Autoflex mass spectrometer from Bruker Daltonics and analysed. The mass spectra of *B. anthracis* exhibited discriminating biomarkers at 4606, 5413 and 6679 Da. Although the data-analysis methods were complex, the *B. anthracis* isolates were correctly classified into two different clusters of six subgroups of the *B. cereus* group.

MALDI-typing of *Listeria monocytogenes*

Listeria monocytogenes is an important food-borne pathogen (Kathariou, 2002), and the differentiation of *L. monocytogenes* from other *Listeria* species and *L. monocytogenes* subtyping are important tools for epidemiological investigations of food-related illnesses. Barbuddhe *et al.* (2008) in their previous study demonstrated that MALDI-TOF MS performed using a Microflex LT instrument accurately identified 146 strains of *Listeria* (representing six species) at the species level and correctly subtyped all strains of *L. monocytogenes*, which was in agreement with PFGE results.

Use of MALDI-typing for Gram-negative bacteria

MALDI-typing of *Salmonella* spp.

Salmonella remains one of the main causes of gastrointestinal and bloodstream infections worldwide, especially in African countries, and the main serovars are *S. enterica* serovar Typhi and non-typhoid *Salmonella*. The current techniques implemented for typing isolates are fastidious and not available in all countries (Wain *et al.*, 2015). In 2012, Kuhns *et al.* evaluated the usefulness of MALDI-TOF for typing 225 *S. enterica* clinical isolates from blood cultures of patients in Ghana along with 44 *S. enterica* reference strains. Although the standard Biotyper software was unable to properly differentiate serovars, it was possible to discriminate clinically important subtypes of *Salmonella* with a detailed analysis, and the authors found at least six peaks that were able to differentiate *S.* Typhi from non-*S.* Typhi isolates (Kuhns *et al.*, 2012).

MALDI-typing of *Escherichia coli*

E. coli represents the most frequently isolated bacterial species in a wide range of human infections, and thus the rapid identification of a pathotype and/or an outbreak is a critical challenge for microbiologists and clinicians. Clark *et al.* were the first to report the successful use of MALDI-TOF MS for typing a series of 136 *E. coli* isolates, representing eight distinct pathotypes (Clark *et al.*, 2013b). To achieve classification, the authors used a combination of the presence and absence of specific peaks for pathotype classification (Clark *et al.*, 2013b). Similarly, Novais *et al.* demonstrate as proof of concept that MALDI-TOF is a good tool for discriminating high-risk *E. coli* clones from phylogenetic groups B2 (ST 131) and D (ST69, ST405, ST393) (Novais *et al.*, 2014). Finally, Christner *et al.* recently showed that MALDI-TOF MS was able to identify shiga-toxigenic *E. coli* isolates during the large outbreak of *E. coli* O104:H4 in Germany in 2011 (Christner *et al.*, 2014). The authors reported two characteristic peaks at m/z 6711 and m/z 10883 that allowed the correct classification of 292 out of 293 isolates, including 104 isolates from this outbreak (Christner *et al.*, 2014). Interestingly, in this study, the classification and discovery of these peaks was found retrospectively from the spectra acquired from routine species identification, showing that this could have been performed in real-time and thus represents an alternative for real-time outbreak surveillance (Christner *et al.*, 2014). Finally, it was recently confirmed that MALDI-TOF can be used as a routine tool to quickly assign *E. coli* isolates to phylogroups A, B1, B2 and D in a large series of 656 well-characterized clinical and environmental isolates (Sauget *et al.*, 2014).

MALDI-typing of *Klebsiella pneumoniae*

K. pneumoniae represents one of the main causes of infections in humans that are associated with nosocomial outbreaks in intensive care units. MALDI-TOF MS has been recently used for the biotyping of a series of 535 *K. pneumoniae* clinical isolates from France and Algeria (Berrazeg *et al.*, 2013). Clustering of spectra allowed the identification of five distinct clusters; data mining revealed that the Algerian isolates clustered together and were associated with respiratory infections and an ESBL phenotype, whereas the isolates from France were more likely to be associated with urinary tract infections and a wild-type phenotype (Berrazeg *et al.*, 2013). Similarly, Bernaschi *et al.* reported the usefulness of a hierarchical approach for typing multidrug-resistant *K. pneumoniae* clinical isolates in a paediatric hospital (Bernaschi *et al.*, 2013).

MALDI-typing of *Acinetobacter baumannii* and *Pseudomonas* spp.

A. baumannii is also a major pathogen responsible for nosocomial infections, and typing these bacteria is critical to implement isolation and control policy measures. MALDI-TOF MS has been used successfully to differentiate outbreak-associated *A. baumannii* clinical isolates from controls and non-outbreak strains (Spinali *et al.*, 2015). Similar results were obtained by Mencacci *et al.* in a series of 35 multidrug-resistant clinical isolates of *A.*

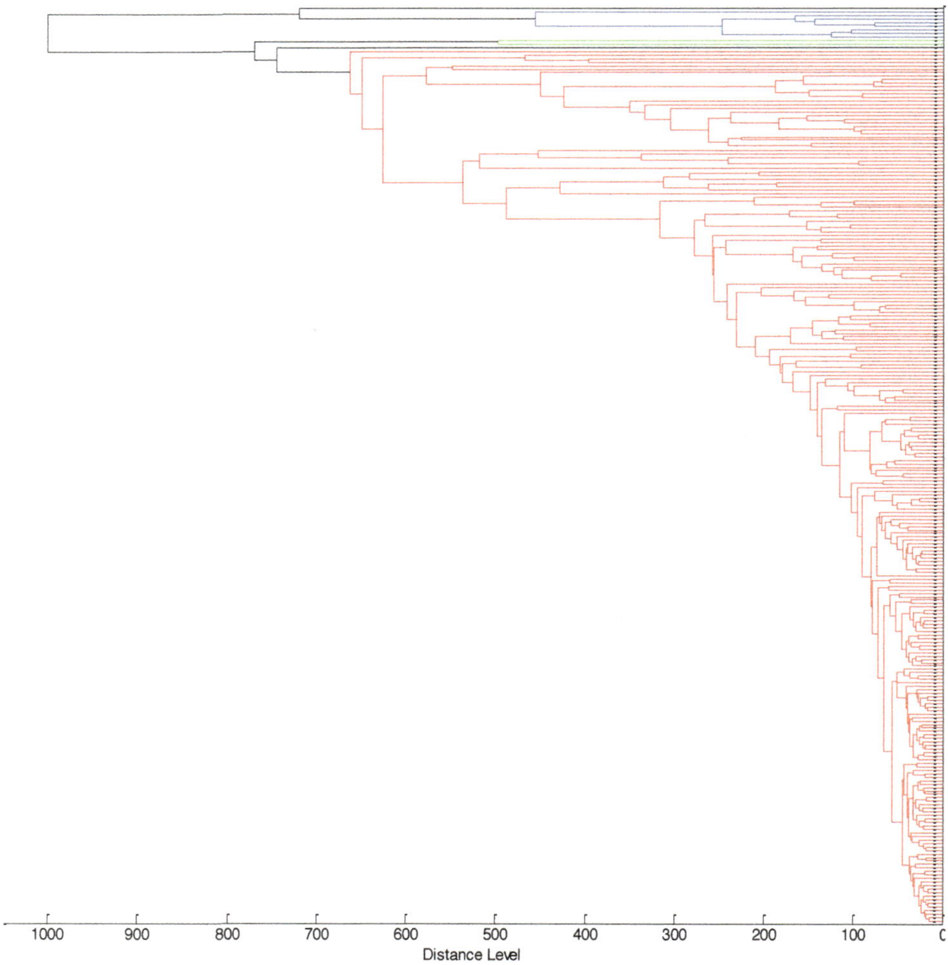

Figure 4.2 Example of an MSP dendrogram showing five distinct clusters of *P. aeruginosa* clinical isolates according to an arbitrary cut-off at the distance level of 500.

baumannii responsible for an outbreak in a general hospital in Italy in 2010, suggesting that MALDI-TOF MS can be used as a routine tool for the real-time detection of nosocomial outbreaks (Mencacci *et al.*, 2013). Finally, MALDI-TOF MS was also successfully applied for typing a series of isolates of *Pseudomonas stutzeri*, including clinical isolates (Scotta *et al.*, 2013). An example of the clustering of *P. aeruginosa* clinical isolates is shown in Fig. 4.2 (unpublished data).

MALDI-typing of other bacteria

MALDI-TOF MS has also been successfully used for typing a series of other bacteria, including *Legionella pneumophila* isolates (Fujinami *et al.*, 2011), *Ochrobactrum anthropi* clinical isolates (Quirino *et al.*, 2014), *Campylobacter jejuni* isolates (Zautner *et al.*, 2013),

Leptospira spp. and serovars (Calderaro *et al.*, 2014), and *Mycoplasma pneumoniae* clinical isolates (Xiao *et al.*, 2014).

Use of MALDI-typing for fungi

MALDI-TOF MS has changed diagnosis in the microbiology laboratory and offers new possibilities for the rapid identification of fungi (Croxatto *et al.*, 2012; Chalupova *et al.*, 2014). Rapid and reliable fungal identification is essential for determining the appropriate antifungal treatment, taking into account the natural resistance of the isolate. Traditionally, the diagnosis of yeast and mould infections has depended on morphological identification and microscopic techniques, such as the colour and shape, the topology of each colony, the cell morphology and the induction of sexual reproduction. The conventional phenotypic technique for identifying yeast includes biochemical methods (Vitek ID YST systems (bioMérieux)); chromogenic agar media are also widely used but are limited to unusual yeasts, and misidentifications occur. These different techniques require expertise and are time-consuming.

MALDI-TOF MS-based identification is being adopted by clinical laboratories for the routine identification of fungi. In fact, MALDI-TOF is widely used for the identification of yeasts (*Candida* and *Cryptococcus neoformans*) in clinical microbiology laboratories (Croxatto *et al.*, 2012). In contrast to bacteria, yeasts first require a formic acid protein extraction prior to MALDI-TOF MS analysis due to the glucan and chitin components of their cell wall (Gorton *et al.*, 2014). Several commercial MALDI-TOF MS instruments have been used and evaluated for routine clinical yeast identification in clinical laboratories. The Microflex Biotyper (Bruker Daltonics) has shown a similar species identification rate of 92%, with Vitek MS (bioMérieux) showing 93%, comparable to the biochemical test rate (94%) (Jamal *et al.*, 2014). However, misidentification occurred when the reference spectrum was not available in the database.

For filamentous fungi (mainly *Aspergillus* and *Penicillium*), the real challenge is to develop an efficient protein extraction method (Lau *et al.*, 2013). This is important for reliable identification, but even more for typing purposes. Various methods of sample preparation have been published (Clark *et al.*, 2013a), and a standardized procedure has been developed (Cassagne *et al.*, 2011). This is most likely the most important step in the identification of filamentous fungi, though pigmentation may also represent a limitation for the ionization of proteins (Bader, 2013). Regardless, there are few reference spectra included in the commercially available databases of filamentous fungi (Clark *et al.*, 2013a). Thus, the development of powerful database libraries needs to be performed. Many teams have developed their own databases, with different performance being published (Lau *et al.*, 2013).

MALDI-TOF MS has revolutionized yeast species identification in microbiological laboratories and has also been suggested for identifying subspecies or types in epidemiological analyses (Spinali *et al.*, 2015). Recently, epidemiological testing of fungal isolates was investigated using MALDI-TOF MS to evaluate *Candida parapsilosis* transmission in neonatal intensive care units (Pulcrano *et al.*, 2012). The dendrogram obtained with the MALDI-TOF technique exhibited good agreement compared with the dendrogram obtained by microsatellite analysis, and the MALDI-TOF dendrogram was obtained in a few minutes. This was also demonstrated for biofilm- and non-biofilm-producing *C. parapsilosis* and *C.*

metapsilosis isolates (Kubesova *et al.*, 2012). Thus, the MALDI-TOF MS typing of strains could represent a rapid and reliable method for strain differentiation in nosocomial infection investigations.

Time- and cost-effectiveness

Cost restrictions, training and quality control requirements and the need for rapid turnaround times make MALDI-TOF MS quite appealing compared to conventional microbial identification methods. Compared with conventional identification methods, MALDI-TOF MS has been shown to confer, in most cases, a significant gain of both technician working time and turnaround time. Although the initial cost of a MALDI-TOF MS instrument can be significant ($180,000 to $250,000), the low reagent costs and high identification rates significantly reduce the per-isolate cost for microbial identification. Indeed, phenotypic identification using modern automated platforms costs at least approximately US$10 per isolate (price for reagents, without labour costs), whereas the reagents required for MS-based identification do not exceed $0.50 (Cherkaoui *et al.*, 2010). Because of the extreme speed and low marginal cost of MALDI-TOF MS, it can improve laboratory efficiency. MALDI-TOF MS-based identification provides faster bacterial species identification than conventional phenotypic identification methods, with equal or better accuracy. One of the major advantages of using MALDI-TOF technology for bacterial identification is the time to result, which is reduced from 24 to 48 hours to less than an hour. This is especially relevant for routine clinical microbiology because most results can be reported 1 day earlier. Seng *et al.* conducted a prospective routine MALDI-TOF MS identification analysis of 1660 bacterial isolates with the Bruker system in parallel with conventional phenotypic bacterial identification (Seng *et al.*, 2009). They estimated that the MALDI-TOF identification required an average time of 6 minutes, for an estimated 70–80% reduced cost compared with the conventional methods of identification. Genetic typing methods like pulsed-field gel electrophoresis (PFGE), or multilocus sequence typing (MLST) have a high discriminatory power, however, these methods are time-consuming and cost intensive.

Limitations of MALDI-typing

One of the limitations of MALDI-typing is the lack of guidelines for data interpretation. The statistical analysis of peak patterns is crucial in evaluating type-specific biomarkers. Another major limitation of MALDI-TOF-based microbial typing is primarily due to the algorithmic methods used to analyse the protein profiles and peak patterns (Spinali *et al.*, 2015). Recently, a few studies have described failures to obtain satisfactory typing results using MALDI-TOF MS. Lasch *et al.* (2014) reported an insufficient discriminatory power for the typing of *Staphylococcus aureus* and *Enterococcus faecium* isolates, and Schirmeister *et al.* (2014) reported a similar problem for *Vibrio cholerae* isolates. Finally, increasing the database of specific pathogens may likely enhance the capability of MALDI-TOF to be used as a surrogate for typing and the identification of nosocomial clones and finally to be used as a routine tool for epidemiology and surveillance in clinical microbiology laboratories.

Concluding remarks

MALDI-typing could be used for broad and prospective typing of a majority of bacterial and fungal isolates in clinical microbiology settings. The rapid identification and typing of microbial pathogens have significant public health and medical implications. The introduction of MALDI-TOF MS technology in clinical laboratories will reduce the time required while improving the accuracy of bacterial identification. MALDI-TOF MS is an emerging technology newly applied to the problem of bacterial species identification and is both a highly accurate and rapid method; accordingly, the development of a MALDI-TOF-fingerprint-based typing scheme would be highly desirable. For microbial subtyping, MALDI-TOF MS represents a new promising alternative approach to other conventional methods such as PFGE and MLSA. Although the MALDI-TOF MS technique has a high accuracy for microbial identification and typing, its performance can be significantly improved when more spectra of appropriate reference strains are added to databases. Indeed, database refinement and enrichment are essential elements of MALDI-TOF MS and will allow the method to increase its power as a prospective tool for typing with regard to bacterial species identification. We believe that if these limitations are addressed, MALDI-TOF MS will in the future replace other methods of microorganism typing and surveillance.

References

Bader, O. (2013). MALDI-TOF-MS-based species identification and typing approaches in medical mycology. Proteomics *13*, 788–799.

Barbuddhe, S.B., Maier, T., Schwarz, G., Kostrzewa, M., Hof, H., Domann, E., Chakraborty, T., and Hain, T. (2008). Rapid identification and typing of listeria species by matrix-assisted laser desorption ionization-time of flight mass spectrometry. Appl. Environ. Microbiol. *74*, 5402–5407.

Bernaschi, P., Del, C.F., Petrucca, A., Argentieri, A., Ciofi Degli, A.M., Ciliento, G., Carletti, M., Muraca, M., Locatelli, F., and Putignani, L. (2013). Microbial tracking of multidrug-resistant *Klebsiella pneumoniae* isolates in a pediatric hospital setting. Int. J. Immunopathol. Pharmacol. *26*, 463–472.

Berrazeg, M., Diene, S.M., Drissi, M., Kempf, M., Richet, H., Landraud, L., and Rolain, J.M. (2013). Biotyping of multidrug-resistant *Klebsiella pneumoniae* clinical isolates from France and Algeria using MALDI-TOF MS. PLoS One *8*, e61428.

Biswas, S., and Rolain, J.M. (2013). Use of MALDI-TOF mass spectrometry for identification of bacteria that are difficult to culture. J. Microbiol. Methods *92*, 14–24.

Bittar, F., Cassagne, C., Bosdure, E., Stremler, N., Dubus, J.C., Sarles, J., Reynaud-Gaubert, M., Raoult, D., and Rolain, J.M. (2010). Outbreak of *Corynebacterium pseudodiphtheriticum* infection in cystic fibrosis patients, France. Emerg. Infect. Dis. *16*, 1231–1236.

Bizzini, A., and Greub, G. (2010). Matrix-assisted laser desorption ionization time-of-flight mass spectrometry, a revolution in clinical microbial identification. Clin. Microbiol. Infect. *16*, 1614–1619.

Boggs, S.R., Cazares, L.H., and Drake, R. (2012). Characterization of a *Staphylococcus aureus* USA300 protein signature using matrix-assisted laser desorption/ionization time-of-flight mass spectrometry. J. Med. Microbiol. *61*, 640–644.

Calderaro, A., Piccolo, G., Gorrini, C., Montecchini, S., Buttrini, M., Rossi, S., Piergianni, M., De, C.F., Arcangeletti, M.C., Chezzi, C., and Medici, M.C. (2014). *Leptospira* species and serovars identified by MALDI-TOF mass spectrometry after database implementation. BMC Res. Notes *7*, 330.

Cassagne, C., Ranque, S., Normand, A.C., Fourquet, P., Thiebault, S., Planard, C., Hendrickx, M., and Piarroux, R. (2011). Mould routine identification in the clinical laboratory by matrix-assisted laser desorption ionization time-of-flight mass spectrometry. PLoS One *6*, e28425.

Chalupova, J., Raus, M., Sedlarova, M., and Sebela, M. (2014). Identification of fungal microorganisms by MALDI-TOF mass spectrometry. Biotechnol. Adv. *32*, 230–241.

Cherkaoui, A., Hibbs, J., Emonet, S., Tangomo, M., Girard, M., Francois, P., and Schrenzel, J. (2010). Comparison of two matrix-assisted laser desorption ionization-time of flight mass spectrometry methods with conventional phenotypic identification for routine identification of bacteria to the species level. J. Clin. Microbiol. *48*, 1169–1175.

Christner, M., Trusch, M., Rohde, H., Kwiatkowski, M., Schluter, H., Wolters, M., Aepfelbacher, M., and Hentschke, M. (2014). Rapid MALDI-TOF mass spectrometry strain typing during a large outbreak of Shiga-Toxigenic *Escherichia coli*. PLoS One *9*, e101924.

Clark, A.E., Kaleta, E.J., Arora, A., and Wolk, D.M. (2013a). Matrix-assisted laser desorption ionization-time of flight mass spectrometry: a fundamental shift in the routine practice of clinical microbiology. Clin. Microbiol. Rev. *26*, 547–603.

Clark, C.G., Kruczkiewicz, P., Guan, C., McCorrister, S.J., Chong, P., Wylie, J., van, C.P., Tabor, H.A., Snarr, P., Gilmour, M.W., Taboada, E.N., and Westmacott, G.R. (2013b). Evaluation of MALDI-TOF mass spectroscopy methods for determination of *Escherichia coli* pathotypes. J. Microbiol. Methods *94*, 180–191.

Cosgrove, S.E., and Carmeli, Y. (2003). The impact of antimicrobial resistance on health and economic outcomes. Clin. Infect. Dis. *36*, 1433–1437.

Croxatto, A., Prod'hom, G., and Greub, G. (2012). Applications of MALDI-TOF mass spectrometry in clinical diagnostic microbiology. FEMS. Microbiol. Rev. *36*, 380–407.

Degand, N., Carbonnelle, E., Dauphin, B., Beretti, J.L., Le, B.M., Sermet-Gaudelus, I., Segonds, C., Berche, P., Nassif, X., and Ferroni, A. (2008). Matrix-assisted laser desorption ionization-time of flight mass spectrometry for identification of nonfermenting gram-negative bacilli isolated from cystic fibrosis patients. J. Clin. Microbiol. *46*, 3361–3367.

Dieckmann, R., Helmuth, R., Erhard, M., and Malorny, B. (2008). Rapid classification and identification of salmonellae at the species and subspecies level using whole-cell MALDI-TOF mass spectrometry. Appl. Environ. Microbiol. *74*, 7767–7778.

Edwards-Jones, V., Claydon, M.A., Evason, D.J., Walker, J., Fox, A.J., and Gordon, D.B. (2000). Rapid discrimination between methicillin-sensitive and methicillin-resistant *Staphylococcus aureus* by intact cell mass spectrometry. J. Med. Microbiol. *49*, 295–300.

Emonet, S., Shah, H.N., Cherkaoui, A., and Schrenzel, J. (2010). Application and use of various mass spectrometry methods in clinical microbiology. Clin. Microbiol. Infect. *16*, 1604–1613.

Enright, M.C., Day, N.P., Davies, C.E., Peacock, S.J., and Spratt, B.G. (2000). Multilocus sequence typing for characterization of methicillin-resistant and methicillin-susceptible clones of *Staphylococcus aureus*. J. Clin. Microbiol. *38*, 1008–1015.

Facklam, R. (2002). What happened to the streptococci: overview of taxonomic and nomenclature changes. Clin. Microbiol. Rev. *15*, 613–630.

Fujinami, Y., Kikkawa, H.S., Kurosaki, Y., Sakurada, K., Yoshino, M., and Yasuda, J. (2011). Rapid discrimination of *Legionella* by matrix-assisted laser desorption ionization time-of-flight mass spectrometry. Microbiol. Res. *166*, 77–86.

Gorton, R.L., Seaton, S., Ramnarain, P., McHugh, T.D., and Kibbler, C.C. (2014). Evaluation of a short, on-plate formic acid extraction method for matrix-assisted laser desorption ionization-time of flight mass spectrometry-based identification of clinically relevant yeast isolates. J. Clin. Microbiol. *52*, 1253–1255.

Harmsen, D., Claus, H., Witte, W., Rothganger, J., Claus, H., Turnwald, D., and Vogel, U. (2003). Typing of methicillin-resistant *Staphylococcus aureus* in a university hospital setting by using novel software for spa repeat determination and database management. J. Clin. Microbiol. *41*, 5442–5448.

Jackson, K.A., Edwards-Jones, V., Sutton, C.W., and Fox, A.J. (2005). Optimisation of intact cell MALDI method for fingerprinting of methicillin-resistant *Staphylococcus aureus*. J. Microbiol. Methods *62*, 273–284.

Jamal, W.Y., Ahmad, S., Khan, Z.U., and Rotimi, V.O. (2014). Comparative evaluation of two matrix-assisted laser desorption/ionization time-of-flight mass spectrometry (MALDI-TOF MS) systems for the identification of clinically significant yeasts. Int. J. Infect. Dis. *26*, 167–170.

Josten, M., Dischinger, J., Szekat, C., Reif, M., Al-Sabti, N., Sahl, H.G., Parcina, M., Bekeredjian-Ding, I., and Bierbaum, G. (2014). Identification of agr-positive methicillin-resistant *Staphylococcus aureus* harbouring the class A mec complex by MALDI-TOF mass spectrometry. Int. J. Med. Microbiol. *304*, 1018–1023.

Josten, M., Reif, M., Szekat, C., Al-Sabti, N., Roemer, T., Sparbier, K., Kostrzewa, M., Rohde, H., Sahl, H.G., and Bierbaum, G. (2013). Analysis of the matrix-assisted laser desorption ionization-time of flight mass spectrum of *Staphylococcus aureus* identifies mutations that allow differentiation of the main clonal lineages. J. Clin. Microbiol. *51*, 1809–1817.

Kathariou, S. (2002). *Listeria monocytogenes* virulence and pathogenicity, a food safety perspective. J. Food. Prot. *65*, 1811–1829.

Kubesova, A., Salplachta, J., Horka, M., Ruzicka, F., and Slais, K. (2012). Candida 'psilosis'--electromigration techniques and MALDI-TOF mass spectrometry for phenotypical discrimination. Analyst *137*, 1937–1943.

Kuhns, M., Zautner, A.E., Rabsch, W., Zimmermann, O., Weig, M., Bader, O., and Gross, U. (2012). Rapid discrimination of *Salmonella enterica* serovar Typhi from other serovars by MALDI-TOF mass spectrometry. PLoS One *7*, e40004.

Lartigue, M.F., Hery-Arnaud, G., Haguenoer, E., Domelier, A.S., Schmit, P.O., Mee-Marquet, N., Lanotte, P., Mereghetti, L., Kostrzewa, M., and Quentin, R. (2009). Identification of *Streptococcus agalactiae* isolates from various phylogenetic lineages by matrix-assisted laser desorption ionization-time of flight mass spectrometry. J. Clin. Microbiol. *47*, 2284–2287.

Lartigue, M.F., Kostrzewa, M., Salloum, M., Haguenoer, E., Hery-Arnaud, G., Domelier, A.S., Stumpf, S., and Quentin, R. (2011). Rapid detection of 'highly virulent' Group B Streptococcus ST-17 and emerging ST-1 clones by MALDI-TOF mass spectrometry. J. Microbiol. Methods *86*, 262–265.

Lasch, P., Beyer, W., Nattermann, H., Stammler, M., Siegbrecht, E., Grunow, R., and Naumann. D. (2009). Identification of *Bacillus anthracis* by using matrix-assisted laser desorption ionization-time of flight mass spectrometry and artificial neural networks. Appl. Environ. Microbiol. *75*, 7229–7242.

Lasch, P., Fleige, C., Stammler, M., Layer, F., Nubel, U., Witte, W., and Werner, G. (2014). Insufficient discriminatory power of MALDI-TOF mass spectrometry for typing of *Enterococcus faecium* and *Staphylococcus aureus* isolates. J. Microbiol. Methods *100*, 58–69.

Lau, A.F., Drake, S.K., Calhoun, L.B., Henderson, C.M., and Zelazny, A.M. (2013). Development of a clinically comprehensive database and a simple procedure for identification of molds from solid media by matrix-assisted laser desorption ionization-time of flight mass spectrometry. J. Clin. Microbiol. *51*, 828–834.

Lowy, F.D. (1998). *Staphylococcus aureus* infections. N. Engl. J. Med. *339*, 520–532.

McDonald, L.C., Killgore, G.E., Thompson, A., Owens, R.C., Jr., Kazakova, S.V., Sambol, S.P., Johnson, S., and Gerding, D.N. (2005). An epidemic, toxin gene-variant strain of *Clostridium difficile*. N. Engl. J. Med. *353*, 2433–2441.

Mellmann, A., Cloud, J., Maier, T., Keckevoet, U., Ramminger, I., Iwen, P., Dunn, J., Hall, G., Wilson, D., Lasala, P., *et al.* (2008). Evaluation of matrix-assisted laser desorption ionization-time-of-flight mass spectrometry in comparison to 16S rRNA gene sequencing for species identification of nonfermenting bacteria. J. Clin. Microbiol. *46*, 1946–1954.

Mencacci, A., Monari, C., Leli, C., Merlini, L., De Carolis, E., Vella, A., Cacioni, M., Buzi, S., Nardelli, E., Bistoni, F., *et al.* (2013). Typing of nosocomial outbreaks of *Acinetobacter baumannii* by use of matrix-assisted laser desorption ionization-time of flight mass spectrometry. J. Clin. Microbiol. *51*, 603–606.

Mitchell, B.G., and Gardner, A. (2012). Mortality and *Clostridium difficile* infection: a review. Antimicrob. Resist. Infect. Control, *1*, 20.

Moura, H., Woolfitt, A.R., Carvalho, M.G., Pavlopoulos, A., Teixeira, L.M., Satten, G.A., and Barr, J.R. (2008). MALDI-TOF mass spectrometry as a tool for differentiation of invasive and noninvasive *Streptococcus pyogenes* isolates. FEMS. Immunol. Med. Microbiol. *53*, 333–342.

Murray, P.R. (2010). Matrix-assisted laser desorption ionization time-of-flight mass spectrometry: usefulness for taxonomy and epidemiology. Clin. Microbiol. Infect. *16*, 1626–1630.

Novais, A., Sousa, C., de Dios, C.J., Fernandez-Olmos, A., Lopes, J., Ramos, H., Coque, T.M., Canton, R., and Peixe, L. (2014). MALDI-TOF mass spectrometry as a tool for the discrimination of high-risk *Escherichia coli* clones from phylogenetic groups B2 (ST131) and D (ST69, ST405, ST393). Eur. J. Clin. Microbiol. Infect. Dis. *33*, 1391–1399.

Pignone, M., Greth, K.M., Cooper, J., Emerson, D., and Tang, J. (2006). Identification of mycobacteria by matrix-assisted laser desorption ionization-time-of-flight mass spectrometry. J. Clin. Microbiol. *44*, 1963–1970.

Pulcrano, G., Roscetto, E., Iula, V.D., Panellis, D., Rossano, F., and Catania, M.R. (2012). MALDI-TOF mass spectrometry and microsatellite markers to evaluate *Candida parapsilosis* transmission in neonatal intensive care units. Eur. J. Clin. Microbiol. Infect. Dis. *31*, 2919–2928.

Quirino, A., Pulcrano, G., Rametti, L., Puccio, R., Marascio, N., Catania, M.R., Matera, G., Liberto, M.C., and Foca, A. (2014). Typing of *Ochrobactrum anthropi* clinical isolates using automated repetitive extragenic palindromic-polymerase chain reaction DNA fingerprinting and matrix-assisted laser desorption/ionization-time-of-flight mass spectrometry. BMC Microbiol. *14*, 74.

Reil, M., Erhard, M., Kuijper, E.J., Kist, M., Zaiss, H., Witte, W., Gruber, H., and Borgmann, S. (2011). Recognition of *Clostridium difficile* PCR-ribotypes 001, 027 and 126/078 using an extended MALDI-TOF MS system. Eur. J. Clin. Microbiol. Infect. Dis. *30*, 1431–1436.

Rizzardi, K., and Åkerlund, T. (2015). High molecular weight typing with MALDI-TOF MS – a novel method for rapid typing of *Clostridium difficile*. PLoS One 10, e0122457.

Sauer, S., and Kliem, M. (2010). Mass spectrometry tools for the classification and identification of bacteria. Nat. Rev. Microbiol. *8*, 74–82.

Sauget, M., Nicolas-Chanoine, M.H., Cabrolier, N., Bertrand, X., and Hocquet, D. (2014). Matrix-assisted laser desorption ionization-time of flight mass spectrometry assigns *Escherichia coli* to the phylogroups A, B1, B2 and D. Int. J. Med. Microbiol. *304*, 977–983.

Schirmeister, F., Dieckmann, R., Bechlars, S., Bier, N., Faruque, S.M., and Strauch, E. (2014). Genetic and phenotypic analysis of *Vibrio cholerae* non-O1, non-O139 isolated from German and Austrian patients. Eur. J. Clin. Microbiol. Infect. Dis. *33*, 767–778.

Scotta, C., Gomila, M., Mulet, M., Lalucat, J., and Garcia-Valdes, E. (2013). Whole-cell MALDI-TOF mass spectrometry and multilocus sequence analysis in the discrimination of *Pseudomonas stutzeri* populations: three novel genomovars. Microb. Ecol. *66*, 522–532.

Seng, P., Drancourt, M., Gouriet, F., La Scola, B., Fournier, P.E., Rolain, J.M., and Raoult, D. (2009). On-going revolution in bacteriology: routine identification by matrix-assisted laser desorption ionization time-of-flight mass spectrometry. Clin. Infect. Dis. *49*, 543–551.

Seng, P., Rolain, J.M., Fournier, P.E., La, S.B., Drancourt, M., and Raoult, D. (2010). MALDI-TOF-mass spectrometry applications in clinical microbiology. Future. Microbiol. *5*, 1733–1754.

Sogawa, K., Watanabe, M., Sato, K., Segawa, S., Ishii, C., Miyabe, A., Murata, S., Saito, T., and Nomura. F. (2011). Use of the MALDI BioTyper system with MALDI-TOF mass spectrometry for rapid identification of microorganisms. Anal. Bioanal. Chem. *400*, 1905–1911.

Sogawa, K., Watanabe, M., Sato, K., Segawa, S., Miyabe, A., Murata, S., Saito, T., and Nomura, F. (2012). Rapid identification of microorganisms by mass spectrometry: improved performance by incorporation of in-house spectral data into a commercial database. Anal. Bioanal. Chem. *403*, 1811–1822.

Spinali, S., van, B.A., Goering, R.V., Girard, V., Welker, M., Van, N.M., Pincus, D.H., Arsac, M., and Durand, G. (2015). Microbial typing by MALDI-TOF MS: Do we need guidance for data interpretation? J. Clin. Microbiol. *53*, 760–765.

Tuller, T., Girshovich, Y., Sella, Y., Kreimer, A., Freilich, S., Kupiec, M., Gophna, U., and Ruppin, E. (2011). Association between translation efficiency and horizontal gene transfer within microbial communities. Nucleic Acids Res. *39*, 4743–4755.

van Veen, S.Q., Claas, E.C., and Kuijper, E.J. (2010). High-throughput identification of bacteria and yeast by matrix-assisted laser desorption ionization-time of flight mass spectrometry in conventional medical microbiology laboratories. J. Clin. Microbiol. *48*, 900–907.

Vargha, M., Takats, Z., Konopka, A., and Nakatsu, C.H. (2006). Optimization of MALDI-TOF MS for strain level differentiation of *Arthrobacter* isolates. J. Microbiol. Methods *66*, 399–409.

Wain, J., Hendriksen, R.S., Mikoleit, M.L., Keddy, K.H., and Ochiai, R.L. (2015). Typhoid fever. Lancet. *385*, 1136–1145.

Walker, J., Fox, A.J., Edwards-Jones, V., and Gordon, D.B. (2002). Intact cell mass spectrometry (ICMS) used to type methicillin-resistant *Staphylococcus aureus*: media effects and inter-laboratory reproducibility. J. Microbiol. Methods *48*, 117–126.

Williamson, Y.M., Moura, H., Woolfitt, A.R., Pirkle, J.L., Barr, J.R., Carvalho, M.G., Ades, E.P., Carlone, G.M., and Sampson, J.S. (2008). Differentiation of *Streptococcus pneumoniae* conjunctivitis outbreak isolates by matrix-assisted laser desorption ionization-time of flight mass spectrometry. Appl. Environ. Microbiol. *74*, 5891–5897.

Wolters, M., Rohde, H., Maier, T., Belmar-Campos, C., Franke, G., Scherpe, S., Aepfelbacher, M., and Christner, M. (2011). MALDI-TOF MS fingerprinting allows for discrimination of major methicillin-resistant *Staphylococcus aureus* lineages. Int. J. Med. Microbiol. *301*, 64–68.

Xiao, D., Zhao, F., Zhang, H., Meng, F., and Zhang, J. (2014). Novel strategy for typing *Mycoplasma pneumoniae* isolates by use of matrix-assisted laser desorption ionization-time of flight mass spectrometry coupled with ClinProTools. J. Clin. Microbiol. *52*, 3038–3043.

Zautner, A.E., Masanta, W.O., Tareen, A.M., Weig, M., Lugert, R., Gross, U., and Bader, O. (2013). Discrimination of multilocus sequence typing-based *Campylobacter jejuni* subgroups by MALDI-TOF mass spectrometry. BMC Microbiol. *13*, 247.

MALDI-TOF MS for Determination of Resistance to Antibiotics

5

Jaroslav Hrabák, Monika Dolejská and
Costas C. Papagiannitsis

Abstract

Antibiotic resistance is a growing problem of current medicine worldwide, having significant impact on human health and economy. Rapid determination of susceptibility to antimicrobials in bacteria including detection of resistance mechanisms are crucial not only for successful treatment of infectious diseases in individual patients, but also for prevention of nosocomial acquisition and further transmission of multidrug resistant bacteria. For that reason, there is an urgent need to develop highly sensitive and automatized methods that can be implemented in microbiology diagnostic laboratories. The current research demonstrates that MALDI-TOF MS is becoming a very promising tool for detection of antibiotic resistance. Several techniques have been developed up to know, some of them, such as MALDI-TOF detection of carbapenemases, have been already introduced into routine diagnostics. Pioneering works using MALDI-TOF for qualitative or quantitative evaluation of susceptibility or resistance as well as several techniques for detection clinically relevant pathogens (e.g. methicillin-resistant *Staphylococcus aureus*, vancomycin-resistant enterococci) or specific antibacterial activity (e.g. β-lactamase, ribosomal RNA methyltransferase) have been published and highlight many pitfalls during the early method development. Introduction of novel techniques using MALDI-TOF MS approaches as well as optimization, validation and full automation of these methods, increasing their reproducibility and decreasing the costs, are urgently needed. It is quite evident that MALDI-TOF MS technology will play an important role in routine diagnostics of antibacterial susceptibility as well as in epidemiological surveillance, overall, improving the quality of health care.

Antibiotic resistance

Infectious diseases are one of the most important issues of medicine. For centuries, therapy of those diseases was limited, resulting in high mortality. In the 19th century, Louis Pasteur, Joseph Lister and Ignaz Semmelweiss carried out pioneering studies, demonstrating successful results of proper prophylaxis of infectious diseases. Causal therapy, however, had been unfeasible for the following decades till the middle of the 20th century (Bryskier, 2005). Since 1936, when sulfonamides were introduced into clinical practice, followed by penicillin in 1941, the face of medicine changed completely, allowing development of many branches, such as surgery, intensive care, transplantology, etc.

In 1889, Paul Vuillemin used the term 'antibiont' for microorganisms that inhibit growth of other microorganisms. In 1941, Selman A. Waksman named all natural substances inhibiting bacterial growth at low concentration as 'antibiotics'. Based on this definition, antibiotics are substances of biological or semisynthetic origin (Bryskier, 2005, EUCAST 2000). Fully synthetic drugs (e.g. quinolones, sulfonamides) should be mentioned as antibacterial agents and not antibiotics (Bryskier, 2005). However, the term 'antibiotics' is currently used also for antimicrobial agents of synthetic origin (EUCAST 2000).

Despite the fact that in the 1950s some people believed that infectious diseases will pass away because of antibiotics, resistance to penicillin mediated by β-lactamases was described shortly after the introduction of penicillin into clinical practice (Kirby, 1944). Based on our current knowledge, no antibiotic can remain stable without developing of resistance in bacteria. In case of some antibiotics, resistance can appear after > 10 years (e.g. glycopeptides, gentamicin, carbapenems), in other classes, immediately after their introduction to the market (e.g. levofloxacin, linezolid, ceftaroline) (Hede, 2014).

Antibiotic resistance is the ability of bacteria or other microorganisms to resist the effects of an antibiotic. Resistance of microorganisms can have two meanings – microbiological and clinical resistance (EUCAST, 2000).

Microbiological resistance is defined as a presence of any resistance mechanism, demonstrated either phenotypically or genotypically. Based on this definition, susceptible bacteria are those that belong to the basic (susceptible) population lacking any resistance mechanism (EUCAST, 2000). Bacterial population can be characterized by a distribution of minimum inhibitory concentration (MIC) or inhibition zone diameters.

Clinical resistance of a microorganism is defined as a high likelihood of therapeutic failure even if maximum doses of a given antibiotic are administered. Clinically susceptible microorganism is the one that ideally responds to a standardized therapeutic regimen when causing an infection (EUCAST, 2000).

Bacterium can be also classified as intermediate susceptible if the outcome of the therapy of an infection caused by such microorganism may vary based on the site of infection and antibiotic dosage.

All of the parameters mentioned above are determined using *in vitro* testing. Currently, specific values such as MICs and inhibition zone diameters are used to setting up epidemiological cut-off values categorizing a bacterium to microbiological resistant or susceptible category. The same parameters are daily used by diagnostic microbiology laboratories to provide information on clinical susceptibility or resistance for specific infections (EUCAST, 2013).

Need for detection of antibiotic resistance and its mechanisms

As mentioned above, antibiotic resistance has been developed after the introduction of any antibiotic to the market. The most important resistant bacteria are multidrug-resistant *Mycobacterium tuberculosis*, vancomycin-resistant *Enterococcus* spp., multiresistant *Enterobacteriaceae* (i.e. *Klebsiella pneumoniae*, *Enterobacter* spp., *Escherichia coli*, etc.), *Pseudomonas aeruginosa* and *Acinetobacter baumannii* (Levy and Marshall, 2004). In developing countries, spectrum of clinically/epidemiologically important pathogens includes *Salmonella enterica*, *Shigella* spp. and *Vibrio cholerae* (Levy and Marshall, 2004).

The epidemiological situation, however, changes with time. In 1990s, methicillin-resistant *Staphylococcus aureus* (MRSA) represented an important issue in European countries and

North America. Infections caused by MRSA showed higher mortality and increased extra economic costs needed for the treatment of infected patients (Levy and Marshall, 2004, de Kraker *et al.*, 2011). Since the first decade of the 2000s, a shift in the burden of antibiotic resistance from Gram-positive to Gram-negative bacteria has been observed (de Kraker *et al.*, 2011). Currently, the most important issue of antibiotic resistance is the increasing resistance to carbapenems, last-line drugs for life-threatening infections (Tzouvelekis *et al.*, 2014). Infections caused by carbapenemase-producing bacteria are difficult to treat with only few antibiotic choices, such as colistin. In some countries, however, *Enterobacteriaceae* strains (especially *K. pneumoniae*) resistant to all available antibiotics have been described (Tóth *et al.*, 2010, Monaco *et al.*, 2014).

Successful reduction of carbapenem-resistant *Enterobacteriaceae* was done in Israel between 2007 and 2012. Due to strict intervention including fast and accurate laboratory diagnosis, and prevention of nosocomial acquisition, a decline from monthly 55.5 cases of patients colonized/infected by carbapenem-resistant *Enterobacteriaceae* per 100,000 patient-days to annual 4.8 cases per 100,000 patient-days has been observed (Schwaber and Carmeli, 2014). For such successful intervention, however, there is an urgent need for diagnostic laboratories to introduce rapid and sensitive methodologies for the detection of carbapenemase-producers (Hrabák *et al.*, 2014).

Automatic tools including MALDI-TOF MS for detection of antibiotic resistance can be used for:

1. rapid determination of MIC equivalent for categorization of the isolate to susceptible/intermediate/resistant category with a direct impact on the patient therapy;
2. detection of resistance mechanism for interpretive reading (Leclercq *et al.*, 2013). In some cases (e.g. MRSA, extended-spectrum β-lactamase (ESβL)/carbapenemase producers), resistance mechanism can predict a treatment outcome better than 'classical' susceptibility testing methods;
3. determination of resistance mechanism for epidemiological purpose (e.g. to isolate colonized or infected patients, to reveal routes of transmission).

Categorization of clinical isolates based on their clinical breakpoints

As mentioned above, classification of clinical isolates to 'susceptible' or 'resistant' category is based on the **minimum inhibitory concentration** or inhibition zone diameters. MIC is defined as the lowest concentration of antibiotic that, under defined *in vitro* conditions, prevents the growth of bacteria within a defined period of time (EUCAST, 2000). This value, however, does not mean that the bacteria are killed by this antibiotic concentration. Apart from MIC, two parameters describing biological effect of an antibiotic on bacteria may be also defined. First, the **minimum bactericidal concentration** (MBC) is the lowest concentration of an antibiotic that, under defined *in vitro* conditions, reduces by 99.9% (3 logarithms) the number of organisms in a medium containing a defined inoculum of bacteria, within a defined period of time (EUCAST, 2000). As mentioned in EUCAST definitions, this effect can be determined by time-kill curve, in which an inoculum is incubated with the antibiotic and samples are tested for numbers of surviving colony forming units (CFUs) at defined time intervals (EUCAST, 2000). Second, the **minimum antibacterial**

concentration (MAC) is a concentration below MIC that can exert specified biological effects on bacteria (e.g. partial inhibition of growth below the MIC, changes in bacterial morphology, changes in adhesion to surfaces) (EUCAST, 2000).

A pioneering work showing a possible use of MALDI-TOF MS to detect changes in a cell after exposure to antimicrobials was published by Marinach *et al.* in 2009. For the experiments, they used *Candida albicans* and fluconazole as a common antifungal agent for treatment of candidiasis. They hypothesized that the protein composition can vary after exposition to different drug concentrations. This observation can be used for prediction of treatment efficiency. In that procedure, microbes are cultivated in standard cultivation media. In early exponential phase, the media are replaced with the media containing antimicrobial agent. After further cultivation, mass spectra of microbes non-exposed and those exposed to antimicrobials are compared. As tested on 17 well-characterized *C. albicans* isolates, their results clearly correlated with MIC values. Because the measured values are rather MAC than MIC, they defined a new parameter – **minimal profile change concentration** (MPCC).

De Carolis *et al.* (2012) showed that the methodology developed by Marinach *et al.* (2009) can be also important for detection of caspofungin susceptibility of various species of *Candida* and *Aspergillus*. They found an excellent correlation between MPCC and MIC (or **minimum effective concentration** (MEC) in case of *Aspergillus* spp.). Turnaround -time of the assay was 15 hours compare with 24 hours needed for standard susceptibility testing. Similar results were obtained using caspofungin susceptible and resistant isolates of *C. albicans* and simplified protocol shortening exposition to a break-point concentration of antifungal agent to 3 hours (Vella *et al.*, 2013).

Saracli *et al.* (2015) validated the assay on large collection of *Candida* spp. and fluconazole, voriconazole, or posaconazole. They found an agreement between MALDI-TOF MS assay and 'classical' susceptibility testing varied in the range of 54% and 97%. As well as the reproducibility of the MALDI-TOF MS assay was low, varying between 54.3% and 82.9%.

Similarly to antifungal agents, the response of bacteria to antibiotics visualized by differences in MALDI-TOF MS spectra can be observed. In 2013, Demirev *et al.* published a comparison of MALDI-TOF MS profiles of *Bacillus anthracis* and *Bacillus globigii* cultivated in standard cultivation medium and isotope-enriched growth medium containing streptomycin. As mentioned by the authors of that study, stable isotopes for labelling of biomolecules of bacteria are commonly used for more than 50 years. If the media contain a predefined isotope ratio (e.g. ^{13}C to ^{12}C and/or ^{15}N to ^{14}N), MALDI-TOF MS can be used for quantitative evaluation of the expression level of cellular proteins (Demirev *et al.*, 2013). In case of resistant isolates, the bacteria continue to grow in isotope-labelled cultivation media. Thus, isotope-labelled nutrients are incorporated into metabolites causing the shift of peaks in mass spectra. In proper (break-point) concentration of antibiotic in cultivation media, MALDI-TOF mass spectra contain different peaks than the spectra obtained from cells cultivated in media without any antibiotic or from susceptible isolates.

Jung *et al.* (2014) validated that assay on meropenem, tobramycin and ciprofloxacin susceptible and resistant *Pseudomonas aeruginosa*. They used a medium supplemented with isotope-labelled lysine. Using an automated algorithm, they found an excellent concordance for all three antibiotics.

The response of bacteria to antibiotics in automated systems is usually detected by spectrophotometric detection of absorbance. Using a proper internal standard (e.g. RNase B), semi-quantitative MALDI-TOF MS with automated interpretation of the spectra can provide comparable results (Lange *et al.*, 2014). After a proper cultivation time in presence of tested antibiotic, spectra of resistant isolates show similar peak intensity as cells cultivated in antibiotic-free media. Spectra of susceptible isolates show weak expression of cellular proteins.

Based on the definition, it should not be expected that those methods can predict MIC as determined by dilution or gradient diffusion tests. The response of bacteria to antibiotics is not a rapid change but specific killing curves with more or less growth continuation are observed depending on certain antibiotic. Therefore, it is better to use the proposed parameter MPCC or already established values such as MAC or MBC. Determination of susceptibility/resistance in bacteriostatic antibiotic may also be difficult.

All of the assays mentioned above may provide useful information for proper antibiotic therapy more rapidly compared to 'classical' susceptibility testing. They can be important in cases when MALDI-TOF MS is integrated in automated lines in diagnostic laboratories. Such tests, if performed automatically, could be done independently with no need of additional equipment (e.g. spectrophotometers). However, further optimization and validation of these methods are needed.

Determination of resistance mechanisms

Direct detection of β-lactamase activity

β-Lactamases are enzymes hydrolysing amide bond of β-lactam antibiotics resulting in an increase of molecular weight by 18 Da. After hydrolysis, the degradation products of some β-lactams are further decarboxylated (–44 Da). β-Lactamases are categorized based on their molecular properties (Ambler classification) or their hydrolytic pattern (Bush classification) (Bush *et al.*, 1995; Bush and Jacoby 2010). Currently, hundred different types of these enzymes have been described (http://www.lahey.org/studies).

Based on their structure of the active site, β-lactamases are categorized to serine enzymes and zinc-dependent metalloenzymes. Metallo-β-lactamases (MBLs) are able to hydrolyse almost all β-lactams (including carbapenems) with some exceptions such as monobactam (aztreonam). The most diverse group of serine β-lactamases includes enzymes with narrow-spectrum activity (e.g. penicillinases) as well as β-lactamases hydrolysing all available β-lactam molecules (Bush and Jacoby, 2010; Bush *et al.*, 1995). Until now, few inhibitors of β-lactamases have been introduced to clinical practice. Unfortunately, there is no common inhibitor for all β-lactamase types.

Infections caused by β-lactamase-producing bacteria are associated with significant mortality, in small clinical studies ranging from 22% to 72% (Borer, 2009; Hirsch and Tam, 2010). The reasons for the increased mortality are numerous, including underlying diseases, delays in the initiation of antibiotic treatment and the lack of effective antibiotics (Patel *et al.*, 2008; Nordmann *et al.*, 2011). The most important β-lactamases are carbapenemases, enzymes inactivating carbapenem antibiotics. Until now, however, no equal molecule to substitute carbapenem antibiotics for treatment of severe infections caused by those bacteria has been introduced. Therefore, preventing the spread of multidrug-resistant bacteria in healthcare settings and the community is a big challenge for current medicine. For this purpose, the introduction of rapid and sensitive assays for the detection of carbapenemase-producing bacteria is of a global importance.

In 2011, two groups (Hrabák *et al.*, 2011; Burckhardt and Zimmermann, 2011) independently demonstrated that MALDI-TOF MS is able to detect changes of β-lactam molecule. Both assays were designed to detect carbapenemase activity using meropenem (Hrabák *et al.*, 2011) or ertapenem (Burckhardt and Zimmermann, 2011) as the indicator molecule. In 2012, Sparbier *et al.*, showed that MALDI-TOF MS is able to detect not only carbapenems (meropenem, ertapenem), but also other β-lactams (e.g. ampicillin, third-generation cephalosporins). Therefore, all β-lactamases [e.g. extended-spectrum β-lactamase- (ESBL) and AmpC-type enzymes] based on their substrate specificity can also be detected.

The principle of the method is very simple – a fresh, usually overnight, bacterial culture is inoculated to a suspension buffer and centrifuged. The pellet is re-suspended in a reaction buffer containing the β-lactam molecule. After incubation at 35°C for 0.5–3 hours, the reaction mixture is centrifuged, and the supernatant is used for MALDI-TOF MS measurement. The α-cyano-4-hydroxycinnamic acid (CHCA) is used as matrix in the majority of published methods. In the assays using meropenem, 2,5-dihydroxybenzoic acid (DHB) seems to be a proper matrix (Fig. 5.1), because the dimer of CHCA (m/z of 380) may cover the peak of sodium salt of decarboxylated degradation product (*m/z* 380.5) or intact meropenem molecule (384.5 *m/z*) (Fig. 5.2). The spectra containing peaks representing the β-lactam

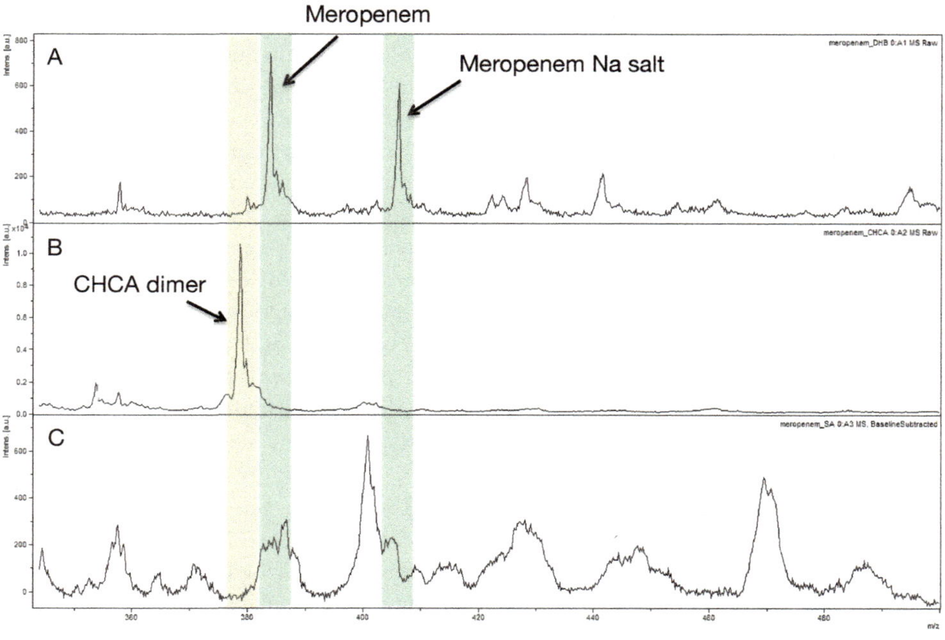

Figure 5.1 Influence of different matrices on detection of β-lactam molecule (meropenem). (A) Using a 2,5-dihydroxybenzoic acid (DHB), meropenem and its sodium salts are clearly visible. (B) The dimer of α-cyano-4-hydroxycinnamic acid (CHCA, *m/z* 379) covers meropenem molecule (*m/z* 384). (C) Mass spectra acquired using 3,5-dimethoxy-4-hydroxycinnamic acid. This matrix does not allow to detect any meropenem molecule or its sodium salt.

molecule, its salts (usually sodium salts) and/or its degradation products are then analysed (Hrabák *et al.*, 2011, 2013).

In the spectra, precise molecular mass of β-lactam and its sodium salts (+22 Da) is detected. Degradation products usually appear as molecules with a mass shift of +18 Da (hydrolysis of amide bound) or −26 Da (decarboxylated hydrolysed form) (Sparbier *et al.*, 2012).

Since the first description of the assay in 2011, many articles describing a modification and validation of MALDI-TOF MS detection of carbapenemases have been published (Kempf *et al.*, 2012; Chong *et al.*, 2015; Papagiannitsis *et al.*, 2015; Studentova *et al.*, 2015; Lasserre *et al.*, 2015).

As demonstrated by Hrabák *et al.* (2012), quality of the spectra may be enhanced using proper reaction conditions (i.e. concentration of bacteria and reaction buffer used). False negative results in some class D carbapenemase (CHDL) producing bacteria (e.g. OXA-48-type enzymes) can be eliminated by an addition of ammonium bicarbonate to the reaction buffer (Papagiannitsis *et al.*, 2015; Studentova *et al.*, 2015). This modification allows a re-carboxylation of active-site lysine of some OXA-type enzymes, resulting in an increased enzymatic activity.

The big challenge for MALDI-TOF MS is a categorization of detected β-lactamase based

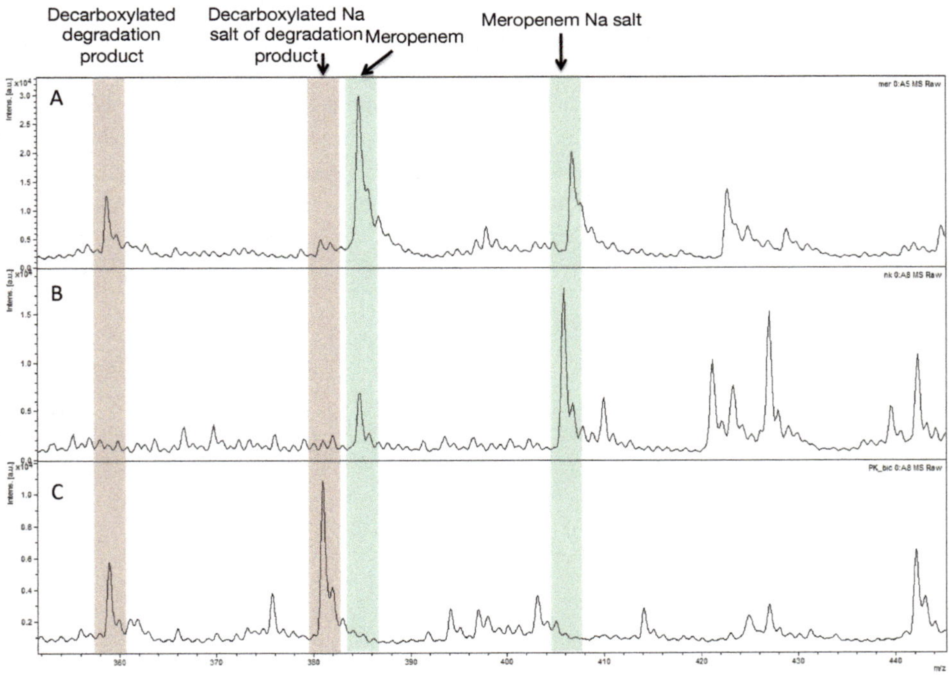

Figure 5.2 Mass spectra acquired by MALDI-TOF MS detection of carbapenemase using meropenem as indicator antibiotic. (A) Spectra of meropenem solution (meropenem *m/z* 384.5, meropenem sodium salt *m/z* 406.5). (B) *Escherichia coli* non-producing any carbapenemase. (C) KPC-2-producing *Klebsiella pneumoniae* (decaboxylated meropenem degradation product *m/z* 358.5, sodium salt of decarboxylated degradation product *m/z* 380.5).

on its inhibitory profile. For ESBLs, serine inhibitors such as clavulanic acid and tazobactam can be easily used. For discrimination of different types of carbapenemases (e.g. EDTA for MBLs, phenylboronic acid for KPCs) inhibitors may, however, interact during MALDI-TOF MS measurement resulting in non-readable spectra.

Recent publications mentioned above have demonstrated excellent sensitivity and specificity of MALDI-TOF MS detection of carbapenemases, allowing to classify this assay as a gold standard (Hrabák *et al.*, 2014). The main disadvantage, however, is the manual interpretation of mass spectra requiring an experienced worker. Therefore, there is an urgent need for software providing automatic reading and interpretation of the spectra. Such a tool has been recently developed by Bruker Daltonics (MBT STAR-BL) and is currently under laboratory evaluation (MBT STAR-BL) (Fig. 5.3). Using proper thresholds and calibrants, the software may provide excellent results (Papagiannitsis *et al.*, 2015).

Direct detection of β-lactamase activity applicable in routine diagnostic microbiology laboratories is crucial for treatment of patients as well as for proper establishment of isolation procedures of infected/colonized patients. This is the only way how to control the spread of carbapenemase-producing bacteria as demonstrated by Schwaber and Carmeli (2014).

The main advantage of MALDI-TOF MS-based detection of carbapenemase activity is a detection of real hydrolysis products clearly proving a disruption of amide bound.

Detection of β-lactamase types

For identification of β-lactamase types, direct visualization of the enzymes based on their molecular mass has been a big challenge for MALDI-TOF MS applications in microbiology (Hrabák *et al.*, 2013). In 2007, Camara and Hays detected a peak corresponding to TEM-1 β-lactamase. They differentiated wild-type *Escherichia coli* (ATCC 700926) from ampicillin-resistant (Amp^R) plasmid-transformed *E. coli* strains by the direct visualization of a β-lactamase. The authors used sinapinic acid as a matrix and noted that this peak was

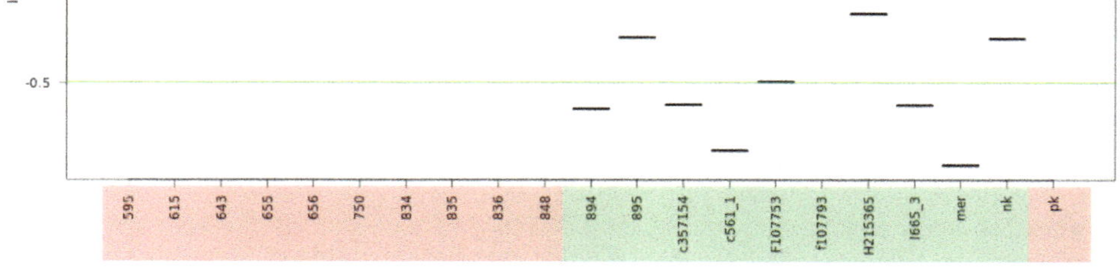

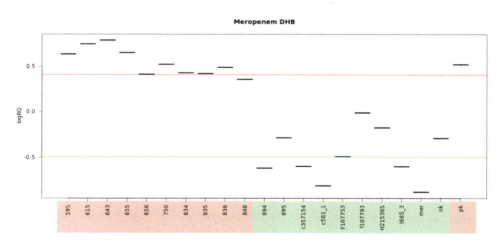

Figure 5.3 Automatic analysis of MALDI-TOF MS detection of carbapenemases. Positive samples are highlighted by red colour, negative samples are highlighted by green colour.

not detectable using whole cells for bacterial identification nor using the other matrices (2,5-dihydroxyacetophenone ferulic acid, 2,5-dihydroxybenzoic acid).

Following the pilot study published by Camara and Hays (2007), Schaumann *et al.* (2012) demonstrated that MALDI-TOF MS is not able to distinguish clinical isolates of *Enterobacteriaceae* and *Pseudomonas aeruginosa* producing either ESβLs or MβLs from non-producers. The measurement, however, ranged from m/z 2000 – 12000, which seems to be smaller than the expected molecular mass of common β-lactamases (> 27 kDa).

First successful assay allowing the detection of CMY-2-like β-lactamases in wild-type isolates was published by Papagiannitsis *et al.* (2014). The assay is based on the extraction of periplasmic proteins and the detection of a peak specific for CMY-2-like β-lactamases using sinapinic acid as a matrix. This method demonstrates that for successful detection it may be important to extract β-lactamases from the periplasmic space to suppress cytoplasmic proteins in spectra. In that assay, β-lactamases were extracted using sucrose-method and stabilized before ethanol precipitation using meropenem as a β-lactamase substrate. Another β-lactamases of the ca. 39,670-m/z, and ca. 38,900-m/z corresponding to ACC-4 and DHA-1 enzymes, were also detected (Papagiannitsis *et al.*, 2014). These results indicate that MALDI-TOF MS may be a powerful tool for discrimination of the diverse groups of AmpC-type β-lactamases. The method was able to detect peaks with an increased m/z of approximately 383, representing the acyl–enzyme complex (complex of CMY-2 β-lactamase with the meropenem molecule) as well.

This method can be used for epidemiological purpose to detect β-lactamase variant in research as well as diagnostic laboratories. MALDI-TOF MS allows a profiling of β-lactamases presented in the isolate like isoelectric focusing of bacterial extracts that is used in many epidemiological studies prior molecular-genetic testing.

Detection of ribosomal RNA methyltransferase activity

Many antibiotic groups (e.g. aminoglycosides, chloramphenicol, clindamycin) inhibit protein synthesis of bacterial cell due to binding to ribosomes. Methylation of the RNA ribosomal subunit confers resistance to these types of drugs (Smith and Baker, 2002). In 2000, Kirpekar *et al.* developed a method for the detection of modifications to ribosomal RNA with MALDI-TOF MS. This method was also used for analysis of the methylation of 23S rRNA by the product of *cfr* gene conferring resistance to chloramphenicol, florfenicol and clindamycin (Kehrenberg *et al.*, 2005). Detection of the methyltransferase activity responsible for the resistance to aminoglycosides caused by the methylation of 16S rRNA was described in 2009 by Savic *et al.*

However, these methods can be used for research purpose only, since the reaction requires the purification of ribosomes and enzymes which is difficult for routine diagnostic laboratories. After incubation in a special buffer, the rRNA is digested with a specific RNase to yield smaller products that can be visualized by MALDI-TOF MS. Methylation of the molecule is detected as an increase of the mass by 14 Da.

MRSA detection

MRSA is a major pathogen both within hospitals and in the community. The mechanism of resistance to methicillin is based on the expression of an additional PBP (PBP2a) that is resistant to the action of the antibiotic but at the same time it can perform the function of the host PBPs, the cross-linking reaction during the cell wall synthesis. MRSA differs genetically from methicillin-sensitive *S. aureus* (MSSA) isolates by the presence the SCC*mec* element in the chromosome, which, among others, contains *mecA* gene encoding the 76 kDa PBP2a. Discrimination between MSSA and MRSA strains is a big challenge for MALDI-TOF MS applications in routine microbiology. In the first study published by Edwards-Jones *et al.* (2000), the authors detected 14 MRSA-specific peaks (e.g. *m/z* of 511, 563, 640, 1165, 1229 and 2127) and two MSSA-specific peaks (*m/z* of 2548, 2647) in the *m/z* ranges of 511 to 2127 and 2548 to 2647, respectively. They used intact bacterial cells and 5-chloro-2-mercapto-benzothiazole as the matrix. Subsequently, Du *et al.* (2002) were also able to discriminate between MRSA and MSSA. Unlike to the first studies mentioned above, Bernardo *et al.* (2002) were not able to detect MRSA isolates despite the fact they used bacterial lysates for their analysis.

Another promising modification of MALDI-TOF MS was a surface-enhanced laser desorption/ionization time-of-flight (SELDI-TOF) MS with selective binding of proteins to a ProteinChip Array. Although Shah *et al.* (2011) were successful in discrimination between MRSA ($n = 49$) and MSSA ($n = 50$) by the detection of seven specific peaks (3081, 5893 and 9580 Da for MSSA, 5709, 7694, 15,308 and 18,896 Da for MRSA), others were not able to reproduce their results.

Because the spread of MRSA is mainly clonal, some of the published studies were focused on detection of specific clones containing *mecA* genes (Ueda *et al.*, 2015). Using that approach, unfortunately, new variants as well as new clones possessing *mecA*/*mecC* genes cannot be detected. In case of local epidemiology; however, such assays can represent fast methods applicable for infection control.

In 2014, M. Josten *et al.* published a study showing a presence of a small peptide – phenol-soluble modulin (PSM) – that is encoded on SCC*mec* cassette I, III and VIII. This protein corresponds to a peak of 2415 *m/z*. The peak representing δ-toxin (*m/z* of 3007) may also help in the detection of MRSA isolates. For the analysis, the measurement should be performed without specific extraction using a standard procedure of the deposition of culture on the MALDI-target covered with α-cyano-4-hydroxy-cinnamic acid as a matrix in 50% acetonitrile/2.5% trifluoroacetic acid solution. Measurement should be performed with a special focus on *m/z* 2411–2419 area. The main disadvantage of this assay is the inability to detect MRSA strains possessing other SCC*mec* cassettes or those encoding the *mecC* gene.

All of the methods for MRSA detection by MALDI-TOF MS that have been described, however, are not able to detect specific penicillin-binding protein 2a (PBP2a). Therefore, their specificity and sensitivity may be limited based on the clonality of analysed isolates.

Detection of vancomycin-resistant *Enterococcus* spp. and *Staphylococcus aureus*

Vancomycin-resistant enterococci (VRE) have emerged as important nosocomial pathogens. Vancomycin prevents the synthesis of peptidoglycan precursors of the bacterial cell wall by blocking the transglycosylation step. The basic mechanism of vancomycin resistance is the formation of peptidoglycan receptors with reduced affinity to the antibiotic. This results in decreased binding of vancomycin and decreased inhibition of cell wall synthesis. Gene clusters corresponding to various vancomycin resistance phenotypes (e.g. vanA, vanB, vanC, vanD, vanL, etc.) has been described.

Griffin *et al.* (2012) demonstrated the ability of MALDI-TOF MS to track a clonal lineage of vancomycin-resistant enterococci. In the study, they used 67 *vanB*-positive *Enterococcus faecium* isolates that were prepared by formic acid extraction method before MALDI-TOF MS measurement. The authors report a sensitivity and specificity of the assay of 96.7% and 98.1%, respectively. They also compared MALDI-TOF MS with PFGE, which is a gold standard technique for epidemiological tracking during an outbreak. Both methods showed an excellent concordance. Relevant results, however, were not obtained with direct plating of the culture on MALDI target due to the thickness of the peptidoglycan in the cell wall, which does not allow proper acquisition of the spectra. In that study, the authors used only strains that were isolated from a single hospital during a short period of 18 months. Other authors were not able to reproduce their results and to identify putative VRE-corresponding peaks (Lasch *et al.*, 2015). Therefore, the results of Griffin *et al.* (2012) seem to be restricted to specific resistant clones only. This discrepancy nicely demonstrates that for such studies, isogenic strains must be used for validation of all methods detecting epidemiological/resistance markers.

Another challenge of current clinical microbiology is the detection of glycopeptide-non-susceptible *Staphylococcus aureus* (GISA, glycopeptide-intermediate *S. aureus*; GRSA, glycopeptide-resistant *S. aureus*).

Before MALDI-TOF MS has been widely introduced into clinical laboratories, Majcherczyk *et al.* (2006) published a method for differentiation between teicoplanin-susceptible and teicoplanin-resistant strains of methicillin-resistant *S. aureus*. They used isogenic strains and analysed them by whole-cell method using 5-chloro-2-mercaptobenzothiazole as the matrix. They found differences in the spectra between ca. 2200 and 2800 m/z. Unfortunately, no further studies have been published until now using MALDI-TOF MS to detect GRSA/GISA. Although the phenomenon is of high clinical importance, there is no simple method to differentiate between susceptible and non-susceptible isolates (Hawkey, 2009; Rong and Leonard, 2010).

Detection of porins

Porins are non-specific channel-forming proteins of outer membrane of Gram-negative bacteria that are responsible for diffusion of hydrophilic molecules (e.g. nutrients) to periplasmic space (Nikaido, 2003). The porins are also responsible for transport of some antibiotics (e.g. β-lactams, quinolones) to their target sites. Therefore, alterations of porins (i.e. structural changes, low expression) along with other resistance mechanisms (e.g. production of β-lactamases, quinolone resistance proteins Qnr) increase the resistance level(Fernandéz and Hancock, 2012).

A common method for porin analysis uses sodium N-lauroyl sarcosinate for semi-specific extraction followed by SDS-PAGE. Cai *et al.* (2012) showed that MALDI-TOF MS can be used for visualization of porins based on their molecular weight. The turnaround-time of the assay can be shortened, since there is no need to do SDS-PAGE followed by gel staining. The method, however, should be extended by precise identification of porins using protein fingerprinting. Unfortunately, it is unlikely that clinical laboratories will be able to introduce this assay to their routine schemes. For the purpose of reference and research laboratories, this method can be an important and reliable tool complementing the detection of resistance mechanisms.

Conclusions and future perspectives

MALDI-TOF MS has become an important tool in microbiological routine diagnostics. Since 2010 an increasing number of tests for the detection of antibiotic resistances has been developed. Among them, MALDI-TOF MS detection of carbapenemases is the most promising method, already used routinely by many diagnostic laboratories and reference centres. This is the only method demonstrating the amide bound hydrolysis of the β-lactam ring, which is currently available for routine detection of carbapenemases. The automatic spectra reading and their interpretation is still urgently needed. It is, however, very cheap method that costs < 1 US$ per reaction with excellent sensitivity and specificity. Together with other direct assays (such as Carba NP) this assay should be mentioned as a gold-standard method. Currently, there is no logic explanation to designate amplification techniques (such as PCR) as a gold standard, as they do not show a real carbapenemase activity and fail to detect new enzymes/variants (Hrabak *et al.*, 2014).

Determination of other resistant bacteria (e.g. MRSA, VRE) based on specific markers will be important for clinical diagnostics. Currently published studies, however, have not yet provided a universal method for detection of such resistance types (i.e. specific detection of PBP 2a, changes in peptidoglycan). For the design of such studies, it is very important to use isogenic strains and compare their mass spectra (Fig. 5.4). By this way only, it is possible to avoid detection of clone-specific markers that are not associated with the antibiotic resistance.

To avoid a detection of clone-specific determinants with no connection to antibiotic resistance, it is necessary to use isogenic strains. In case of single gene/determinant responsible for resistance (e.g. β-lactamases, PBP 2a – MRSA, *van* – VRE), the susceptible strain should be transformed using a proper vector. In case of complex resistance mechanisms (e.g. efflux), knock out of proper gene(s) should be performed. This procedure is optimal for single or non-complex resistance mechanisms such as MRSA, VRE, β-lactamase producers.

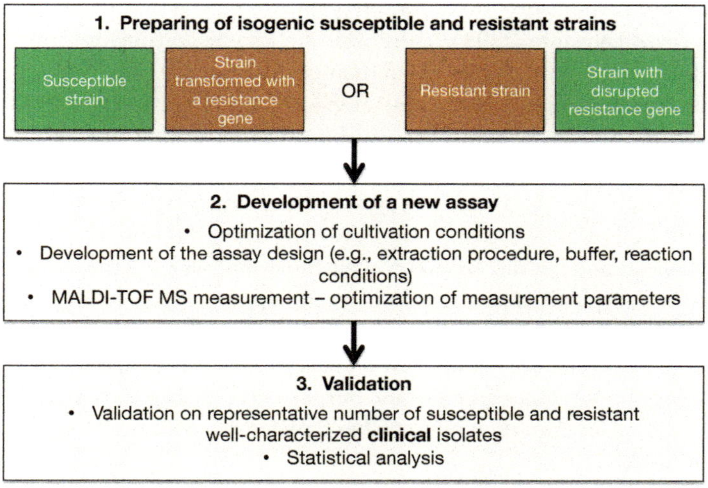

Figure 5.4 Process for development of new MALDI-TOF MS assays for detection of specific resistance determinants.

In case of categorization of clinical isolates to susceptible/intermediate/resistant categories, some promising techniques have also been developed. Their use in routine laboratories is, however, limited because they are still time-consuming and not significantly shorten the time to the result compared with 'classical' antibiotic susceptibility testing. Those methods could play an important role in automation of diagnostic microbiology. It is expected that MALDI-TOF MS will be included in automatic lines. Automation process, perhaps by artificial intelligence algorithms, may significantly shorten the process for obtaining clinically relevant microbiological results.

The use of MALDI-TOF MS for detection of epidemiological markers of antibiotic resistance (e.g. β-lactamases) is evident. We believe that in the future MALDI-TOF MS will play an important role in the process of accelerating susceptibility testing.

Acknowledgements

This work was partially supported by the National Sustainability Program I (NPU I) Nr. LO1503 provided by the Ministry of Education Youth and Sports of the Czech Republic.

References

Bernardo, K., Pakulat, N., Macht, M., Krut, O., Seifert, H., Fleer, S., Hünger, F., and Krönke, M. (2002). Identification and discrimination of *Staphylococcus aureus* strains using matrix-assisted laser desorption/ionization-time of flight mass spectrometry. Proteomics 2, 747– 753.

Borer, A., Saidel-Odes, L., and Riesenberg, K. (2009). Attributable mortality rate for carbapenem-resistant *Klebsiella pneumoniae* bacteremia. Infect. Control Hosp. Epidemiol. *30*, 972–976.

Bryskier, A. (2005). Antimicrobial agents: antibacterial and antifungal. (ASM Press, Washington D.C., USA).

Burckhardt, I., and Zimmermann, S. (2011). Using matrix-assisted laser desorption ionization–time of flight mass spectrometry to detect carbapenem resistance within 1 to 2.5 hours. J. Clin Microbiol *49*, 3321–3324.

Bush, K., and Jacoby, G.A. (2010). Updated functional classification of beta-lactamases. Antimicrob. Agents Chemother. *54*, 969–976.

Bush, K., Jacoby, G.A., and Medeiros, A.A. (1995). A functional classification scheme for beta-lactamases and its correlation with molecular structure. Antimicrob. Agents Chemother. *39*, 1211–1233.

Cai, J.C., Hu, Y.Y., Zhang, R., Zhou, H.W., and Chen, G.X. (2012). Detection of OmpK36 porin loss in *Klebsiella* spp. by matrix-assisted laser desorption ionization-time of flight mass spectrometry. J. Clin. Microbiol. *50*, 2179–2182.

Camara, J.E., and Hays, F.A. (2007). Discrimination between wild-type and ampicillin-resistant *Escherichia coli* by matrix-assisted laser desorption/ionization time-of-flight mass spectrometry. Anal. Bioanal. Chem. *389*, 1633–1638.

De Carolis, E., Vella, A., Florio, A.R., Posteraro, P., Perlin, D.S., Sanguinetti, M., and Posteraro, B. (2012). Use of matrix-assisted laser desorption ionization-time of flight mass spectrometry for caspofungin susceptibility testing of *Candida* and *Aspergillus* species. J. Clin. Microbiol. *50*, 2479–2483.

Chong, P.M., McCorrister, S.J., Unger, M.S., Boyd, D.A., Mulvey, M.R., and Westmacott, G.R. (2015). MALDI-TOF MS detection of carbapenemase activity in clinical isolates of *Enterobacteriaceae* spp., *Pseudomonas aeruginosa*, and *Acinetobacter baumannii* compared against the Carba-NP assay. J. Microbiol. Methods *111*, 21–23.

Demirev, P.A., Hagan, N.S., Antoine, M.D., Lin, J.S., and Feldman, A.B. (2013). Establishing drug resistance in microorganisms by mass spectrometry. J. Am. Soc. Mass Spectrom. *24*, 1194–1201.

Du, Z., Yang, R., Guo, Z., Song, Y., and Wang, J. (2002). Identification of *Staphylococcus aureus* and determination of its methicillin resistance by matrix-assisted laser desorption/ionizationtime-of-flight mass spectrometry. Anal. Chem. *74*, 5487–5491.

Edwards-Jones, V., Claydon, M.A., Evason, D.J., Walker, J., Fox, A.J., and Gordon, D.B. (2000). Rapid discrimination between methicillin-sensitive and methicillin-resistant *Staphylococcus aureus* by intact cell mass spectrometry. J. Med. Microbiol. *49*, 295–300.

European Committee for Antimicrobial Susceptibility Testing (EUCAST) of the European Society of Clinical Microbiology and Infectious Diseases (ESCMID). (2000). Terminology relating to methods for the determination of susceptibility of bacteria to antimicrobial agents. Clinical Microbiology and Infection. *6*, 503–508.

European Committee for Antimicrobial Susceptibility Testing (EUCAST). (2013). Setting breakpoints for new antimicrobial agents. EUCAST SOP 1.1. Available at http://www.eucast.org/documents/sops/

Fernández, L., and Hancock, R.E. (2012). Adaptive and mutational resistance: role of porins and efflux pumps in drug resistance. Clin Microbiol Rev. *25*, 661–681.

Griffin, P.M., Price, G.R., Schooneveldt, J.M., Schlebusch, S., Tilse, M.H., Urbanski, T., Hamilton, B., and Venter, D. (2012). Use of matrix-assisted laser desorption ionization-time of flight mass spectrometry to identify vancomycin-resistant enterococci and investigate the epidemiology of an outbreak. J. Clin. Microbiol. *50*, 2918–2931.

Hawkey, P.M. (2009). Low-level glycopeptide resistance in methicillin-resistant *Staphylococcus aureus* and how to test it. Clin. Microbiol. Infect. *15* (Suppl. 7), 2–9.

Hede, K. (2014). Antibiotic resistance: An infectious arms race. Nature *509*, S2–3.

Hirsch, E.B., and Tam, V.H. (2010). Detection and treatment options for *Klebsiella pneumoniae* carbapenemases (KPCs): an emerging cause of multidrug-resistant infection. J. Antimicrob. Chemother. *65*, 1119–1125.

Hrabák, J., Chudáčková, E., and Papagiannitsis, C.C. (2014). Detection of carbapenemases in *Enterobacteriaceae*: a challenge for diagnostic microbiological laboratories. Clin. Microbiol. Infect. *20*, 839–853.

Hrabák, J., Chudackova, E., and Walkova, R. (2013). Matrix-Assisted Laser Desorption Ionization–Time of Flight (MALDI-TOF) Mass Spectrometry for detection of antibiotic resistance mechanisms: from research to routine diagnosis. Clin. Microbiol. Rev. *26*, 103–114.

Hrabák, J., Studentová, V., Walková, R., Zemlicková, H., Jakubu, V., Chudácková, E., Gniadkowski, M., Pfeifer, Y., Perry, J.D., Wilkinson, K., *et al.* (2012). Detection of NDM-1, VIM-1, KPC, OXA-48, and OXA-162 carbapenemases by MALDI-TOF mass spectrometry. J. Clin. Microbiol. *50*, 2441–2443.

Hrabák, J., Walková, R., Študentová, V., Chudáčková, E., and Bergerová, T. (2011). Carbapenemase Activity Detection by Matrix-Assisted Laser Desorption/Ionisation Time-of-Flight Mass Spectrometry. J. Clin. Microbiol. *49*, 3222–3227.

Josten, M., Dischinger, J., Szekat, C., Reif, M., Al-Sabti, N., Sahl, H.G., Parcina, M., Bekeredjian-Ding, I., and Bierbaum, G. (2014). Identification of agr-positive methicillin-resistant *Staphylococcus aureus* harbouring the class A *mec* complex by MALDI-TOF mass spectrometry. Int. J. Med. Microbiol. *304*, 1018–1023.

Jung, J.S., Eberl, T., Sparbier, K., Lange, C., Kostrzewa, M., Schubert, S., and Wieser, A. (2014). Rapid detection of antibiotic resistance based on mass spectrometry and stable isotopes. Eur. J. Clin. Microbiol. Infect. Dis. 33, 949–955.

Kehrenberg, C., Schwarz, S., Jacobsen, L., Hansen, L.H., and Vester, B. (2005). A new mechanism for chloramphenicol, florfenicol and clindamycin resistance: methylation of 23S ribosomal RNA at A2503. Mol. Microbiol. 57, 1064–1073.

Kempf, M., Bakour, S., Flaudrops, C., Berrazeg, M., Brunel, J.M., Drissi, M., Mesli, E., Touati, A., and Rolain, J.M. (2012). Rapid detection of carbapenem resistance in *Acinetobacter baumannii* using matrix-assisted laser desorption ionization-time of flight mass spectrometry. PLoS One 7, e31676.

Kirby, W.M.M. (1944). Extraction of a highly potent penicillin inactivator from penicillin resistant staphylococci. Science 99, 452–453.

Kirpekar, F., Douthwaite, S., and Roepstorff, P. (2000). Mapping posttranscriptional modifications in 5S ribosomal RNA by MALDI mass spectrometry. RNA 6, 296–306.

de Kraker, M.E., Davey, P.G., Grundmann, H., and BURDEN study group. (2011). Mortality and hospital stay associated with resistant *Staphylococcus aureus* and *Escherichia coli* bacteremia: estimating the burden of antibiotic resistance in Europe. PLoS Med. *8*, e1001104.

Lange, C., Schubert, S., Jung, J., Kostrzewa, M., and Sparbier, K. (2014). Quantitative matrix-assisted laser desorption ionization-time of flight mass spectrometry for rapid resistance detection. J. Clin. Microbiol. *52*, 4155–4162.

Lasch, P., Fleige, C., Stämmler, M., Layer, F., Nübel, U., Witte, W., and Werner, G. (2014). Insufficient discriminatory power of MALDI-TOF mass spectrometry for typing of Enterococcus faecium and *Staphylococcus aureus* isolates. J. Microbiol. Methods *100*, 58–69.

Lasserre, C., De Saint Martin, L., Cuzon, G., Bogaerts, P., Lamar, E., Glupczynski, Y., Naas, T., and Tandé, D. (2015). Efficient detection of carbapenemase activity in *Enterobacteriaceae* by Matrix-Assisted Laser

Desorption Ionization-Time of Flight Mass Spectrometry in less than 30 minutes. J. Clin. Microbiol. *53*, 2163–2171.

Leclercq, R., Cantón, R., Brown, D.F., Giske, C.G., Heisig, P., MacGowan, A.P., Mouton, J.W., Nordmann, P., Rodloff, A.C., Rossolini, G.M., *et al.* (2013). EUCAST expert rules in antimicrobial susceptibility testing. Clin. Microbiol. Infect. *19*, 141–160.

Levy, S.B., and Marshall, B. (2004). Antibacterial resistance worldwide: causes, challenges and responses. Nature Medicine *10*, S122–S129.

Majcherczyk, P.A., McKenna, T., Moreillon, P., and Vaudaux, P. (2006). The discriminatory power of MALDI-TOF mass spectrometry to differentiate between isogenic teicoplanin-susceptible and teicoplanin-resistant strains of methicillin-resistant *Staphylococcus aureus*. FEMS Microbiol. Lett. *255*, 233–239.

Marinach, C., Alanio, A., Palous, M., Kwasek, S., Fekkar, A., Brossas, J.Y., Brun, S., Snounou, G., Hennequin, C., Sanglard, D., *et al.* (2009). MALDI-TOF MS-based drug susceptibility testing of pathogens: the example of *Candida albicans* and fluconazole. Proteomics *9*, 4627–4631.

Monaco, M., Giani, T., Raffone, M., Arena, F., Garcia-Fernandez A., Pollini, S., Network EuSCAPE-Italy, Grundmann, H., Pantosti, A., and Rossolini, G.M. (2014). Colistin resistance superimposed to endemic carbapenem-resistant *Klebsiella pneumoniae*: a rapidly evolving problem in Italy, November 2013 to April 2014. Euro. Surveill. *19*, 20939.

Nikaido, H. (2003). Molecular basis of bacterial outer membrane permeability revisited. Microbiol. Mol. Biol. Rev. *67*, 593–656.

Nordmann, P., Naas, T., and Poirel, L. (2011). Global spread of carbapenemase-producing *Enterobacteriaceae*. Emerg. Infect. Dis. *17*, 1791–1798.

Papagiannitsis, C.C., Kotsakis, S.D., Tuma, Z., Gniadkowski, M., Miriagou, V., and Hrabak, J. (2014). Identification of CMY-2-type cephalosporinases in clinical isolates of *Enterobacteriaceae* by MALDI-TOF MS. Antimicrobial Agents and Chemotherapy *58*, 2952–2957.

Papagiannitsis, C.C., Študentová, V., Izdebski, R., Oikonomou, O., Pfeifer, Y., Petinaki, E., and Hrabák, J. (2015). MALDI-TOF MS meropenem hydrolysis assay with NH_4HCO_3, a reliable tool for the direct detection of carbapenemase activity. J. Clin. Microbiol. *53*, 1731–1735.

Patel, G., Huprikar, S., Factor, S.H., Jenkins, S.G., and Calfee, D.P. (2008). Outcomes of carbapenem-resistant *Klebsiella pneumoniae* infection and the impact of antimicrobial and adjunctive therapies. Infect. Control Hosp. Epidemiol. *29*, 1099–1106.

Rong, S.L., and Leonard, S.N. (2010). Heterogeneous vancomycin resistance in *Staphylococcus aureus*: a review of epidemiology, diagnosis, and clinical significance. Ann. Pharmacother. *44*, 844–850.

Saracli, M.A., Fothergill, A.W., Sutton, D.A., and Wiederhold, N.P. (2015). Detection of triazole resistance among *Candida* species by matrix-assisted laser desorption/ionization-time of flight mass spectrometry (MALDI-TOF MS). Med Mycol. *53*, 736–742.

Savic, M., Lovric, J., Tomic, T.I., Vasiljevic, B., and Conn, G.L. (2009). Determination of the target nucleosides for members of two families of 16S rRNA methyltransferases that confer resistance to partially overlapping groups of aminoglycoside antibiotics. Nucleic Acid Res. *37*, 5420–5431.

Schaumann, R., Knoop, N., Genzel, G.H., Losensky, K., Rosenkranz, C., Stîngu, C.S., Schellenberger, W., Rodloff, A.C., and Eschrich, K. (2012). A step towards the discrimination of beta-lactamase-producing clinical isolates of *Enterobacteriaceae* and *Pseudomonas aeruginosa* by MALDI-TOF mass spectrometry. Med. Sci. Monit. *18*, MT71–77.

Schwaber, M.J., and Carmeli, Y. (2014). An ongoing national intervention to contain the spread of carbapenem-resistant *Enterobacteriaceae*. Clin. Infect. Dis. *58*, 697–703.

Shah, H.N., Rajakaruna, L., Ball, G., Misra, R., Al-Shahib, A., Fang, M., and Gharbia, S.E. (2011). Tracing the transition of methicillin resistance in sub-populations of *Staphylococcus aureus*, using SELDI-TOF mass spectrometry and artificial neural network analysis. Syst. App. Microbiol. *34*, 81–86.

Smith, C.A., and Baker, E.N. (2002). Aminoglycoside antibiotic resistance by enzymatic deactivation. Curr. Drug Targets Infect. Disord. *2*, 143–160.

Sparbier, K., Schubert, S., Weller, U., Boogen, C., and Kostrzewa, M. (2012). Matrix-assisted laser desorption ionization–time of flight mass spectrometry-based functional assay for rapid detection of resistance against beta-lactam antibiotics. J. Clin. Microbiol. *50*, 927–937.

Studentova, V., Papagiannitsis, C.C., Izdebski, R., Pfeifer, Y., Chudackova, E., Bergerova, T., Gniadkowski, M., and Hrabak, J. (2015). Detection of OXA-48-type carbapenemase-producing *Enterobacteriaceae* in diagnostic laboratories can be enhanced by addition of bicarbonates to cultivation media or reaction buffers. Folia Microbiologica *60*, 119–129.

Tóth, A., Damjanova, I., Puskás, E., Jánvári, L., Farkas, M., Dobák, A., Böröcz, K., and Pászti, J. (2010). Emergence of a colistin-resistant KPC-2-producing *Klebsiella pneumoniae* ST258 clone in Hungary. Eur. J. Clin. Microbiol. Infect. Dis. *29*, 765–769.

Tzouvelekis, L.S., Markogiannakis, A., Piperaki, E., Souli, M., and Daikos, G.L. (2014). Treating infections caused by carbapenemase-producing *Enterobacteriaceae*. Clin. Microbiol. Infect. *20*, 862–872.

Ueda, O., Tanaka, S., Nagasawa, Z., Hanaki, H., Shobuike, T., and Miyamoto, H. (2015). Development of a novel matrix-assisted laser desorption/ionization time-of-flight mass spectrum (MALDI-TOF-MS)-based typing method to identify meticillin-resistant *Staphylococcus aureus* clones. J. Hosp. Infect. *90*, 147–155.

Vella, A., De Carolis, E., Vaccaro, L., Posteraro, P., Perlin, D.S., Kostrzewa, M., Posteraro, B., and Sanguinetti, M. (2013). Rapid antifungal susceptibility testing by matrix-assisted laser desorption ionization-time of flight mass spectrometry analysis. J. Clin. Microbiol. *51*, 2964–2969.

Application of MALDI-TOF MS in Veterinary and Food Microbiology

6

Claudia Hess, Merima Alispahic and Michael Hess

Abstract

Considering the huge potential of the technology MALDI-TOF MS in veterinary and food microbiology is still in its infancy. Characterization of bacteria with traditional methods in comparison to genetic analysis is still the predominant task performed in routine diagnostics. In this context, the reliable identification of pathogenic bacteria is of high relevance and the application of MALDI-TOF MS is a valuable addition to existing technologies. Furthermore, proteomic data have gained access to resolve taxonomy/phylogeny. In food microbiology the detection of zoonotic bacteria has a high priority and MALDI-TOF MS can already be applied on major food-borne bacteria, e.g. *Campylobacter* spp., *Salmonella* spp., *Listeria* spp. and *E. coli*. The diverse functions of lactobacilli and acetic acid bacteria, being essential for the fermentation process towards the spoilage of food products, prompted studies on these microorganisms. This resulted in the transformation of knowledge obtained from single bacteria to a microbial community in a certain product or production process.

Independent of the organism and the targeted subject the establishment of biomarkers has a high priority for easy discrimination and identification. The increasing application of MALDI-TOF MS in routine veterinary or food microbiology asks for further development and extension of the technique which will be addressed in a separate chapter.

Introduction

In veterinary and food microbiology the identification of a strain to species, and sometimes subspecies, or even variant level is required. Without accurate identification, estimations about prevalence and significance of different species in a given environment are not possible. Veterinary clinical management and food safety is regularly facilitated if the identity of a strain is known.

Veterinary microbiology is still dominated by traditional methods used to identify bacteria, like growth requirements, cell morphology, biochemical and immunological tests. Phenotypic tests remain a commonly used approach to identify bacteria, however inconsistency of phenotypic profiles and similarities of phenotypic characters, prevent accurate identification. Identification involves the comparison of data obtained for an unknown strain with those of known taxa. An isolate is identified when phenotypic (e.g. biochemical tests, fatty acid or protein profiles) or genotypic (e.g. DNA fingerprinting, sequencing and PCR techniques) data matches those determined for a defined taxon to an acceptable level.

Food safety is of increasing concern as the consumers prefer wide-ranging of foods and beverages that are not only traded nationally but also globally. Food spoilage is caused by microorganisms, enzymes and chemical action, with bacteria being the major cause of spoilage of most food products. Thus, spoilage microbiota includes both spoilage and non-spoilage bacteria, often making it difficult to determine the true nature of the bacteria present. Most bacterial species that are related to food deterioration are able to form volatile metabolites like histamine, ammonia, organic acids, hypoxanthine, acetate, trimethylamine and volatile compounds of sulfur, resulting in off-flavours and even intoxication. Despite this, numerous food products rely on the presence of certain microorganisms for the fermentation process and other chemical reactions to obtain the final product.

A vast number of methods have been developed for the identification of genera, species, and strains of bacteria with relevance in veterinary microbiology. These consist of culture methods (biochemical tests and chemical analysis of pure cultures), serological methods (serotyping) and genetic methods (nucleic acid hybridization, sequencing and PCR techniques). The common methods for identifying bacteria in contaminated samples involve pure cultures, obtained in a first step by streaking or sequential dilution. Microscopic observation and biochemical tests are followed by verification of the suspected identity with specific antibodies or molecular probes and PCR, which are for some bacteria commercially available. DNA microarrays can be used to test for numerous species of bacteria at one time. Nevertheless, if none of the probes correspond to any of the bacteria in the sample, then other tests with pure cultures will be necessary. Each of the methods has advantages for identification of certain species; however, none has yet accomplished all of the advantages of speed and accuracy without certain limitations.

MALDI-TOF MS has just recently approached veterinary and food microbiology. Together with new sequence technologies the fast and rapid identification of bacteria has been substantially improved in recent years. Considering the work flow of daily veterinary clinical or food microbiology MALDI-TOF MS offers numerous advantages, although various applications and challenges are still waiting to be addressed.

Veterinary microbiology

General considerations

The benefit of MALDI-TOF MS as diagnostic tool in veterinary microbiology is mainly appreciated as a method providing a high throughput of samples within a short period of time. Traditional routine microbiological methods still rely on phenotypic identification like colony growth of bacteria on various culture media, morphological and biochemical characterization of the isolated bacteria, altogether laborious and time-consuming. Furthermore, it is known that some bacteria are difficult to classify based upon phenotypic identification due to variable species characteristics. In some cases even DNA based methods are unable to resolve all ambiguities. The advantages of MALDI-TOF MS in this context are very obvious and are already used for different applications. In recent years MALDI-TOF MS was introduced as one of the most adaptable chemotaxonomic methods in veterinary medicine to address questions with regard to phylogeny and taxonomy, which will be in the focus of the next chapter.

Diagnosing a bacterial infection is based on the correct and reliable identification of the causative agent as outlined in the second section within the chapter on veterinary microbiology. This is crucial to implement efficient treatment and a prerequisite to reconstitute animal health. The possible transmission of bacteria to other animal species and even humans emphasizes the importance for unambiguous identification. The detection of protein profiles by MALDI-TOF MS became a convenient tool for rapid analysis of bacteria and was already implemented in routine diagnostics of veterinary laboratories. This is of special importance considering the huge number of microorganisms inhabiting animals, sometimes restricted to a specific animal species. The fact that numerous bacteria can be transmitted from animals to humans – and sometimes vice versa – underlines the blurred border between microorganisms addressed in this chapter.

Characterization of bacteria with relevance for veterinary medicine in a taxonomic or phylogenomic context

The need for a reliable up-to-date reference database based on well-defined reference and field strains is very obvious to address questions with regard to taxonomy and phylogeny. Considering the high number of animal pathogens and animal species together with a high priority of the existing databases on human medicine the lack of well characterized reference strains isolated from animals is very obvious and supported by comparing genetic, biochemical and proteomic characterization (Wragg *et al.*, 2014). Consequently, establishing a suitable database is done by individual laboratories as part of their ongoing research. A common approach is to compare different traditional and genetic methods with newly established MALDI-TOF MS and to determine the degree of overlap.

Species of the family *Pasteurellaceae* play an important role as primary or opportunistic animal pathogens. The identification of this group of bacteria in veterinary diagnostic laboratories is mainly done by phenotypic assays. MALDI-TOF MS seems to represent a promising alternative compared with currently practised phenotypic diagnostics. A very comprehensive study investigated 250 strains comprising 15 genera and more than 40 species and subspecies, altogether covering most representatives of the family (Kuhnert *et al.*, 2012). MALDI-TOF MS was shown to represent a highly potent method for the diagnosis of this group of animal pathogens characterized by a high discrimination power at the genus and species level. Similar results were noticed by investigating 66 reference strains belonging to the four recognized species, *G. anatis*, *G. melopsittaci*, *G. salpingitidis* and *G. trehalosifermentans*, in the genus *Gallibacterium* (Alispahic *et al.*, 2011). These bacteria represent a phenotypically heterogeneous group, which complicates species identification when traditional methods are applied (Christensen *et al.*, 2003). By implementing MALDI-TOF MS clear species identification can be achieved based on the unique spectra obtained from each of the four *Gallibacterium* species with 87.6% concordance to biochemical/physiological characterization. Discrimination of strains from different farms or geographic regions is also possible (Alispahic *et al.*, 2012; Paudel *et al.*, 2013). Obtained data indicate that clonal lineages can be established from isolates taken from different farms, emphasizing a certain clustering of subspecies, a pattern noticed by AFLP as well. Robustness of the findings was independent of the tested MALDI-TOF MS instruments and strains can be stored up to 8 days at 4°C without influence on the quality of spectra. Discrimination is less good for bacterial species of the genus *Avibacterium* within the *Pasteurellaceae*, despite reproducible spectra of good quality (Alispahic *et al.*, 2014). Obtained data also showed differences for genetic typing based upon multilocus sequence phylognetic analysis. It was concluded that the failure of proteomic characterization is due to the fact that the investigated *Avibacterium* spp. are less diverge and therefore might be not distinct species at all.

Similar problems are faced with regard to unquestionable characterization of members of the genera *Brachyspira* causing diarrhoea in pigs and poultry. Species identification is often performed by applying different PCR protocols as well as a number of DNA-based typing techniques. Applying MALDI-TOF MS to investigate *Brachyspira* spp. reference strains revealed unique protein profiles for each species (Calderaro *et al.*, 2013; Prohaska *et al.*, 2014). These results indicate that MALDI-TOF MS facilitates the diagnosis of swine dysentery and porcine intestinal spirochaetosis with unique peaks noticed for the different bacterial species. Application on field isolates confirmed reliability in comparison to routinely performed sequencing of the *nox* gene (Prohaska *et al.*, 2014).

Corynebacterium species are among the most frequently isolated pathogens associated with subclinical mastitis in dairy cows. Conventional tests to identify *Corynebacterium* spp. are characterized by a high misidentification rate (Coyle and Lipsky, 1990). Investigating 180 *Corynebacterium* strains isolated from milk samples from cows with intramammary infections resulted in 96.1% correct and reliable identifications to genus and species level by MALDI-TOF MS (Goncalves *et al.*, 2014). This outcome leads to the conclusion that the method could serve as a potent alternative to achieve species-level diagnoses of bovine intramammary infections caused by *Corynebacterium* spp.

In case of bovine mastitis *Enterococcus faecalis* and *Enterococcus faecium* are also frequently isolated (Werner *et al.*, 2012). Species misidentification by biochemical and phenotypic assays are known, which were handled by applying sequence analysis. MALDI-TOF MS was compared with phenotypic identification methods and PCR to identify 199 *Enterococcus* isolates obtained from epidemiologically unlinked clinically significant bovine mastitis cases. MALDI-TOF MS proved as superior compared with the other methods based on the quick and easy to perform extended direct transfer protocol, with which high-quality spectra were gained.

The genus *Trueperella* was recently assigned to comprise bacteria with similar chemotaxonomic and phylogenetic features of formerly members of the genus *Arcanobacterium* (Yassin *et al.*, 2011). Separation of these two genera is supported by MALDI-TOF MS which shows comparable discrimination similar to genetic methods, like phylogenetic relationship of ribosomal genes (Hijazin *et al.*, 2012a). Bacteria belonging to both genera can be isolated from a great variety of animals, like ruminants, horses, dogs, rabbits and reptiles. *T. abortisuis* could be isolated in the context of abortion in pigs and such isolates cluster differently from other species within the genus by MALDI-TOF MS and 16S rRNA sequencing (Metzner *et al.*, 2013). Similarly, *T. bernadinae* was identified from an anal swab of a piglet showing clinical signs of enteritis (Hijazin *et al.*, 2012b). The fact that this bacteria was detected only recently for the first time from animals asks for further approval with regard to the clinical importance and pathogenicity but also to optimize the conditions used for isolation. In comparison, *Arcanobacterium pluranimalum* is known for some time and can be isolated from ruminants and dogs with very diverse clinical pathologies. In agreement with *Trueperella* spp. differentiation between *A. pluranimalum*, *A. canis* or *A. haemolyticum* from other species within the genus is possible by MALDI-TOF MS similar to 16S rRNA sequencing (Balbutskaya *et al.*, 2014; Sammra *et al.*, 2013, 2014).

The genus *Riemerella* within the *Flavobacteriaceae* is named according to O. Riemer who described in 1904 an exudative septicaemia in geese caused by *R. anatipestifer*. The bacteria can be isolated from various bird species but waterfowl are by far the most important host. Other species in the genus are *R. columbina* and *R. columbipharyngis*. Morphological and biochemical characterization are commonly used but altogether less reliable and serotyping to differentiate between *R. anatipestifer* isolates is hampered by cross-reactions. Instead, MALDI-TOF MS could differentiate between the different species and even a certain subclustering of strains originating from the same geographic area was noticed (Philipp *et al.*, 2013; Rubbenstroth *et al.*, 2011, 2013b). Not only could *R. anatipestifer* be separated by MALDI-TOF from other species, it was also possible to separate *R. columbina* from *R. columbipharyngis*, both isolated from pigeons (Rubbenstroth *et al.*, 2013a). However, applying direct smear preparations to identify *R. columbipharyngis* colonies the bacteria should not be stored as this limits the quality of the spectra. Furthermore, the fact that spectra from only a very limited number of reference strains of *R. anatipestifer* are in the database could limit the precise detection of clinical isolates (Hess *et al.*, 2013).

The lack of reliable genetic methods and the common deficiencies with traditional methods supports application of MALDI-TOF MS to identify staphylococci, which are also involved in cow mastitis, especially coagulase-negative organisms. Using PCR-RFLP based upon the *groEL* gene a good overlap with MALDI-TOF MS can be obtained (Tomazi *et al.*, 2014). Although all isolates can be identified on genus level, misidentification on species level is influenced by the absence of subspecies in the database or insufficient protein extraction method (see below). However, coagulase-positive staphylococci isolated from dogs and cats were successfully characterized following direct transfer of bacteria to the target plate (Decristophoris *et al.*, 2011). Again, good correlation with genetic data were noticed but solid identification of reference strains is needed to prevent insufficient characterization.

Mycoplasmas are bacteria lacking cell walls, carrying minimal genomes and comprising several species, which are important pathogens in veterinary medicine. The identification of mycoplasmas is challenging as phenotypic methods cannot always achieve identification to the species level. So far, the commercial databases contain only a limited number of mycoplasmal MALDI-TOF MS reference spectra, making identification difficult. By constructing a *Mycoplasma* spectral database with ruminant as well as rodent *Mycoplasma* species identification of clinical isolates has been shown (Goto *et al.*, 2012; Pereyre *et al.*, 2013). In general, the recommended scores for genus- and species-level identifications are ≥ 1.700 and ≥ 2.000, respectively. In case of *Mycoplasma* species a reduction of the acceptable scores for species identification to ≥ 1.700 is discussed.

Some species within the genera *Francisella, Leptospira, Mycobacteria* and *Clostridia* are good examples for bacteria with high relevance in veterinary and human medicine. With regard to veterinary medicine diseases like tularaemia in hares, leptospirosis in cattle and swine, tuberculosis in various animal species and black quarter disease caused by *Clostridium chauvoei* are altogether regulated on national or international (Office International des Epizooties (OIE)) level. Following establishment of a reference database the benefit of MALDI-TOF MS to characterize individual species was demonstrated for animal isolates of all these genera. A total of 45 *Francisella* strains were correctly identified with high log score values following establishment of a database, although spectra for *Francisella* spp. were very similar (Seibold *et al.*, 2010). For this reason it is advisable to adapt the pattern recognition algorithm in order to focus on less conserved peaks. Altogether, subspecies of *Francisella tularensis* could be identified, which is of high relevance due to differences in pathogenicity, animal species affected and clinical outcome. Consensus spectra for 19 *Leptospira* species reference isolates were developed directly from freshly grown cultures and used to establish a database (Djelouadji *et al.*, 2012). As only 100 cells are needed to obtain relevant spectra time for propagation can be limited. For *Mycobacteria* spp. it was shown that the time of cultivation influences the spectra (Mather *et al.*, 2014). Following propagation, inactivation of the organisms prior to any manipulation is obligatory to reduce the risk for the personnel. Denaturation of cells by either heat or ethanol can be applied to extract proteins in order to increase the number of peaks/spectra. With MALDI-TOF MS it was possible to separate *Clostridium chauvoi* from *C. septicum*, both sharing 98.5% nucleotide identity within the 16S rRNA, which is of high practical importance due to the applied legislation (Grosse-Herrenthey *et al.*, 2008). Finally, *C. botulinum* isolates could be split into four known metabolic groups, but it was not possible to determine whether an isolate is a neurotoxin producing strain. Although spectra from spores can be used to differentiate *Clostridia* all strains to be tested should be tested under the same conditions to avoid any influence of sporulation.

Confirmation of pathogenic bacteria from specified clinical cases

Traditional methods are still the first choice to characterize isolated bacteria within routine diagnostics as outlined above. In this context MALDI-TOF MS is mainly used to confirm obtained data. Exceptional to this, only MALDI-TOF MS was used to investigate bacteria in stallion samples following aerobic cultivation on Columbia blood agar (Masarikova *et al.*, 2014). A wide spectrum of bacteria from environment to commensal microorganisms was found, with numerous bacteria isolated from more than one animal. Even though MALDI-TOF MS might be used as stand-alone technology confirmation will require the addition of other techniques. Genetic analysis is usually performed to compare samples from a specific clinical situation with proteomic data. *Rhodococcus equi* is a pathogen causing primarily bronchopneumonia in young horses but could also be isolated from swine or cattle with suppurative infections. Following phenotypic characterization isolates were investigated by MALDI-TOF MS and results were compared with sequence analyses of 16S rRNA, *rpoB* and four different *R. equi* genes, some of them plasmid-located. Isolates were unambiguously identified based upon 16S rRNA, whereas *rpoB* sequences varied (Rzewuska *et al.*, 2014). More variations were found comparing presence of plasmid located genes. Applying an extraction step prior to fixation on the steel target plate enhanced reliable identification by MALDI-TOF MS, although not all isolates could be identified.

In agreement with 16S rRNA analysis *Edwardsiella tarda* from a live stranded whale with sepsis was identified correctly by MALDI-TOF MS (Cools *et al.*, 2013; Sammra *et al.*, 2014).

American foulbrood is a notifiable disease in Europe and North America and also listed by the OIE, due to its devastating effect on the brood of honey bees. *Paenibacillus larvae*, a Gram-positive, spore-forming bacterium is the aetiological agent of the disease. Applying ERIC-PCR four genotypes could be assigned, which also differ phenotypically and biochemically. Variations in virulence, driven by the expression of different virulence factors, underline the great heterogeneity of the species. Following ethanol/formic protein extraction high log score values (≥ 2.469) were obtained for the majority of strains isolated worldwide, which were used to set up a reference database (Schäfer *et al.*, 2014). Interestingly, proteomic analysis harmonizes very well with genotyping, especially for ERIC types I and II, the predominant types in European apiaries with worldwide appearance.

Food microbiology

General considerations

In industrialized countries issues of food safety have overtaken food security, emphasizing the need for effective control measures. As a consequence, prevention of zoonotic microbes in food products has a high priority and MALDI-TOF MS was applied in different studies to characterize such bacteria. This should enable to prevent transmission of zoonotic pathogens, the dissemination of resistance genes, virulence factors or any other hazard. It is a common expectation from the consumer that food products are free of any contaminants. In fact food products are mainly regarded as sterile. This assumption is certainly not true except special procedures are implemented post harvesting, like pasteurization. However, this procedure is not applicable for the majority of meat products and other foods asking for special control procedures. This becomes even more evident with the tendency to buy fresh and unprocessed products. Technologies like vacuum or modified atmosphere packing create a certain environment for bacteria asking for in-process control procedures. On the opposite, some foodstuffs need bacteria for fermentation in order to obtain the final product. Independent of the production procedure and the type of food microbiological safety of such products has a high priority.

Characterization of zoonotic bacteria

Owing to their high importance as pathogens transmitted mainly by food *Enterobacteriaceae* are within the focus of technologies like MALDI-TOF MS, enabling rapid identification of these bacteria. Also some of them, like *Campylobacter* or *Salmonella,* induce only limited pathogenicity in animals – if at all – they are of high importance due to the fact that they are leading causes of bacterial enteritis in humans. Precise identification and characterization is needed and biomarkers can be established, altogether helpful to increase diagnostic standards. In this respect MALDI-TOF MS seems favourable in comparison to traditional methods and some re-classification of *Campylobacter* strains was reported in several studies (Kolinska *et al.,* 2008; Mandrell and Wachtel, 1999; Mandrell *et al.,* 2005). Detection of mixed infections of *Campylobacter* species is also possible and harmonizes with data obtained by immunological and biochemical methods (Mandrell *et al.,* 2005). Distinct spectra can be obtained comparing *Helicobacter pullorum, Arcobacter butzleri* with *Campylobacter* spp., together known as food-borne pathogens (Alispahic *et al.,* 2010). Considering the widespread of *C. jejuni* and the existence of several *Campylobater* species, biomarkers offer an additional option to increase specificity and the potential of differentiation. Such biomarkers were mainly allocated in the range between 5 and 15 kDa, whereas ions with a much higher m/z ratio were found less appropriate (Alispahic *et al.,* 2010; Mandrell *et al.,* 2005; Winkler *et al.,* 1999). Various proteins are useful biomarkers but the quality of MALDI-TOF MS spectra and the option to link those with sequence data obtained from *C. jejuni, C. coli, C. upsaliensis* and *C. helveticus* favours a DNA-binding protein (HU) (Fagerquist *et al.,* 2005). The intra- and interspecies variation between this and other proteins, the high copy numbers and four lysine residues at the C-terminus of the HU protein, which suppose to play a role in the strong ionization, altogether supports the usage of this protein as biomarker. Finally, variations of about 15–60 Da are based upon different non-synonymous mutations in amino acid sequences and not due to post-translational modifications, which are conserved between single *C. jejuni* strains (Fagerquist *et al.,* 2005, 2006). This can also be demonstrated for extracted and HPLC separated biomarkers and offers new options to determine the amino acid sequence by proteomic identification without *de novo* MS/MS analysis (Fagerquist *et al.*, 2006, 2007; Fagerquist, 2007). The application of MALDI-TOF MS in clinical studies to type and differentiate *Campylobacter* isolates from animal species is a straightforward process and suitable for routine investigations (Jay-Russell *et al.*, 2012; Klein *et al.*, 2013).

In addition to *Campylobacter*, *Salmonella* are also important food-borne pathogens and nearly all isolates from human salmonellosis belong to the species *Salmonella enterica* subsp. *enterica*. The presence of more than 2500 serotypes underlines the huge diversity of the species. Consequently, discrimination below species level needs spectra with a high number of reproducible protein peaks, which could be achieved by adding sinapinic acid to the bacteria on the target plate (Dieckmann *et al.*, 2008). However, bioinformatics had to be applied to make use of sequence variations within the ribosomal house-keeping proteins for phylogenetic classification. For this, available sequence data are compared with experimental mass data, a procedure omitting less standardized and complex mass spectrometry fingerprint data. Combining biomarkers of all levels from genus to serotype helps to establish a classification algorithm for rapid identification of the most frequently isolated *Salmonella enterica* serovars. This procedure can be used to trace epidemiologically important serovars, namely Enteritidis, Typhimurium, Infantis, Virchow and Hadar (Dieckmann and Malorny, 2011).

Gram-positive *Listeria monocytogenes* is the causative agent of human listeriosis, an important disease in pregnant women and immunocompromised hosts. A variety of meat, dairy and vegetable products act as source for human infections. In veterinary medicine *L. monocytogenes* can be involved in subclinical mastitis in cows and MALDI-TOF MS can be applied to demonstrate the benefit in comparison to biochemical identification (Barreiro *et al.*, 2010). MALDI-TOF MS is also able to discriminate *L. monocytogenes* strains to the same level as PFGE and clonal lineages were established (Barbuddhe *et al.*, 2008). Differentiation of *L. monocytogenes* to other species of the genus *Listeria* is possible based upon spectra within the m/z ratio of 4000–10,000.

In order to optimize the quality of spectra special attention has to be paid towards the media and growing conditions used for typing of food-borne bacteria, because they might influence the expression of certain genes as demonstrated for *Salmonella* (Dieckmann and Malorny, 2011). TSB medium was found most suitable to type *E. coli* (Mazzeo *et al.*, 2006). Campylosel or brucella agar proved to be ideal to investigate *Campylobacter* spp. by MALDI-TOF MS whereas modified charcoal cefoperazone deoxycholate agar, which is often used for initial isolation, is not suitable (Alispahic *et al.*, 2010; Mandrell *et al.*, 2005). Storage of non-fermentable bacteria, such as *Campylobacter*, at 4°C is possible for several days, enabling the collection of isolates to be measured within the same run. As *L. monocytogenes* detection depends on the number of cells in the sample, a medium for enrichment should be used to promote growth, especially in environments harbouring different bacteria (Jadhav *et al.*, 2014). Following this approach 10 cfu of *L. monocytogenes* can be detected and typed from matrixes like milk, pâte and cheese.

Applying MALDI-TOF MS to characterize bacteria from food products

Bacteria play a crucial role in numerous food products and beverages, for example as starter culture for the fermentation process. On the other hand, some bacteria might act as spoilage depending on the food itself, the production process and processing technology. Finally, some bacteria transmitted by food are pathogens in animals and humans with subclinical or clinical appearance.

Milk is one of the most important foods and different techniques are applied post harvesting to reduce or exclude microbiological contamination as raw milk might contain both, zoonotic but also probiotic bacteria. Lysing cells without destroying cell walls followed by inactivation of bacteria with ethanol could be combined with the MALDI Sepsityper Kit™, known from human medicine, to identify *E. coli*, *E. faecalis* and *S. aureus*, which were used to spike pasteurized and whole milk (Barreiro *et al.*, 2012). Incubation for 4 hours might be used in case low numbers of bacteria are present as sensitivity varies by bacterial species. Combining capillary isoelectric focusing with MALDI-TOF MS can be used for unambiguously identification of *Lactobacillus* species in milk samples, but cultivation of bacteria is needed and the fat content has to be < 1.5% (Horka *et al.*, 2013). Although pasteurization is widely practised to sterilize milk, spoilage bacteria are of special concern as excreted proteins, like proteases and lipases, are not destroyed during this process. Different studies showed that MALDI-TOF MS is valuable to identify Gram-positive spoilage bacteria, like *Bacillus* spp.. Moreover, intraspecies information within *Lactobacillus mesenteroides* or *Staphylococcus aureus* can be obtained compared with 16S rRNA analysis (Benmechernene *et al.*, 2014; Böhme *et al.*, 2012; Vithanage *et al.*, 2014). Such comparisons depend on the type of bacteria as for example a higher accuracy can be achieved by 16S rRNA sequencing if Gram-negative *Pseudomonas* spp. are investigated. *S. aureus* is not only a major food-borne pathogen but also a leading cause of mastitis in cows, with high economic losses (see above). MALDI-TOF MS was applied to determine species within the genus *Staphylococcus*, to resolve mixed cultures and discrepancies with classical biochemical procedures (Barreiro *et al.*, 2010; Böhme *et al.*, 2012). Quality of spectra could be improved by processing soluble proteins obtained from cells lysed with an organic solvent (acetonitrile) and a strong acid (aqueous trifluoracetic acid), which might help to identify the bacteria down to the strain level. However, discrimination between methicillin-resistant and -sensitive strains was not possible. Performing an epidemiological study a prevalence of 48.2% of methicillin-resistant coagulase-negative staphylococci was identified from various animal sources and humans (Huber *et al.*, 2011).

A similar procedure as described above should be applied to identify and to establish biomarkers for other Gram-positive bacteria from seafood, like *Bacillus* spp., *Staphylococcus* spp., *Clostridia* spp. and *Listeria* spp. This limits the impact of the matrix and is helpful to standardize the whole process independent of the bacteria, as demonstrated for fish pathogens like *Pseudomonas* spp. or *Aeromonas* spp. (Böhme *et al.*, 2010a,b). Extending the range of available typing methods is of special importance in this area as aquatic farming procedures and processing of seafood harbours the risk of food-borne or spoilage bacteria. For such bacteria MALDI-TOF MS is a valuable addition to genomic typing usually performed by 16S rRNA sequencing and for *Bacillus* spp. and *Pseudomonas* spp. MALDI-TOF MS was found to be more discriminative than genetic analysis on inter- and intraspecies level (Böhme *et al.*, 2013; Fernandez-No *et al.*, 2013). Extending this comparison towards other bacteria demonstrates that bacteria of the same genus cluster together, which is different within the phyloproteomic tree. The proteomic data demonstrate that a certain differentiation between food-borne pathogens from less pathogenic or non-pathogenic bacterial species within a certain genus is feasible as demonstrated for *L. monocytogenes* and *S. aureus* (Böhme *et al.*, 2011).

Various bacteria, e.g. *Pseudomonas* spp. or *Morganella* spp., isolated from farmed fish might be involved in the production of poisoned substances, such as histamine. The identified bacteria differed from other bacterial species present in the reference data base with specific mass peaks in the range of 2534–9113 *m/z* (Fernandez-No *et al.*, 2011). Similar to 16S rRNA sequencing MALDI-TOF MS allowed separation of Gram-positive bacteria from Gram-negative ones, in which *Enterobateriaceae* built a separate cluster (Fernandez-No *et al.*, 2010). Identification of multidrug resistant *E. coli* from minced meat could be achieved following enrichment and isolation of bacteria on selective agar (Petternel *et al.*, 2014). Comparison of MALDI-TOF MS is usually performed with genetic methods as outlined above. Fourier transform infrared (FTIR) spectroscopy is another bioanalytical method, which can be used for identification and differentiation of food bacteria. Whereas ribosomal proteins are targeted by MALDI-TOF MS the FTIR spectra reflect a whole-cell fingerprint based upon much more biochemical data. As a consequence, MALDI-TOF MS is more robust, reliable and less influenced by the growing conditions, which makes it more suitable for daily work. On the opposite, the additional data supplied by FTIR increases the sensitivity and could be used for subspecies identification (Wenning *et al.*, 2014).

Lactic acid and acetic acid bacteria used as probiotics in fermentation, or as spoilage bacteria

Lactobacilli and acetic acid bacteria have a wide spectrum of application and they are widely distributed in animals and food products. They are known as probiotics, starter culture, food contaminants and even spoilage bacteria. For example, beside their potential as bio-preservatives cultures for fermented food products (e.g. cheese), *Enterococcus* species appear usually as secondary contaminants of food products or food spoilage. A peak at *m/z* 4426 was noticed as a common feature of species within the genus *Enterococcus* but whole fingerprints are recommended for species identification (Quintela-Baluja *et al.*, 2013; Favaro *et al.*, 2014). Numerous *Lactobacillus* species exist and MALDI-TOF MS is suitable to identify strains correctly with a probability of < 93%, which is somewhat superior in comparison to genetic methods depending on the applied technology (Duskova *et al.*, 2012). Isolates from chickens and calves could be grouped with high log scores and clusters within certain species could be defined, which needs further approval (Bujnakova *et al.*, 2014). Comparing isolates from geese by MALDI-TOF MS and ITS-PCR and ITS-PCR-RFLP confirmed the ability to separate species by all methods, although intraspecies differentiation could only be achieved using the PCR based techniques (Dec *et al.*, 2014). Strains of *Lactobacillus* species, *Enterococcus* species and *Bifidobacterium animalis* subsp. *lactis* are commonly used as probiotics in fermented milk products. These bacteria are a major component of the microbial barrier to infection and can produce antimicrobial agents. Using MALDI-TOF MS beside molecular fingerprinting methods and sequencing *B. animalis* spp. *lactis* could successfully be identified from dog faeces (Bunesova *et al.*, 2012). Whereas identification of Lactobacilli by MALDI-TOF MS is usually done with whole bacterial cells, determination of biomarkers is more efficient with cell lysates or ribosomal protein fractions (Sun *et al.*, 2006; Teramoto *et al.*, 2007). In this way peak mass obtained for several proteins of *L. bulgaricus* can be compared with available sequence data in the SwissProt database, which might allow allocating sequence errors. Improvement of spectra was achieved by washing cells with ethanol combined with formic acid treatment as demonstrated for *Tetragenococcus halophilus* (Kuda *et al.*, 2014).

Culture conditions of lactobacilli have only less influence on the spectra and bacteria can directly be smeared on the target plate followed by addition of the matrix. For some lactobacilli, like *L. brevis,* showing a different cell wall composition, it is advisable to use cell suspension and not cell extracts (Kern *et al.,* 2013). Lactic acid and acetic acid bacteria are of high relevance as spoilage bacteria in beer and MALDI-TOF MS is very suitable to address the needs of the brewing industry as changes of the bacterial community during the fermentation process can be assessed. Moreover, a certain relationship between Gram-negative *Pectinatus* spp. originating from the same brewery could be found (Kern *et al.,* 2014b; Spitaels *et al.,* 2014). Spectra of acetic acid bacteria might be influenced by the growing conditions with consequences on strain level detection (Wieme *et al.,* 2014a). Considering the different genera and species of bacteria application of similarity coefficients, e.g. the Dice coefficient, on mass peaks was found helpful in order to determine relationship between isolated bacteria (Kern *et al.,* 2014a; Wieme *et al.,* 2014b). Typing of bacteria can be improved by including sequencing analysis of protein encoding genes, which shows a higher taxonomic resolution than 16S rRNA analysis. Other genome analysis techniques like RAPD or DGGE are also widely used to compare proteomic with genetic data.

Conclusions and perspectives

MALDI-TOF MS has gained access into various areas of veterinary and food microbiology, although application so far is mainly restricted to resolve taxonomic ambiguities and to identify bacteria in a certain clinical situation. In this context the method is already a valuable tool in veterinary microbiological diagnostics, extending traditional and genetically based techniques, following a step-wise approach. In numerous cases it was successfully demonstrated that methods used so far are more laborious and the majority could be replaced by applying MALDI-TOF MS. Nevertheless, a broader use can be envisaged supported by various arguments. The speed and minimal costs of sample preparation and measurement for this method makes it exceptionally well suited for routine and high-throughput use. Following evaluation and optimization of validated protocols, reliable and reproducible data can be obtained. The storage and exchange of generated data can easily be arranged between laboratories and research groups. This is more and more needed considering the requirements laid down by regulations with relevance for certification or accreditation of laboratories. Therefore, in order to fulfil such regulations standardized protocols would be needed in order to allow comparisons between individual laboratories. Interlaboratory studies as a routine basis for diagnosticians using MALDI-TOF MS, a procedure well known in other areas, would complete such efforts and increase the confidence in the method. This would have severe implications not only for identification of bacteria from clinical samples or foods, but also for the characterization of bacteria supposed to be included in autogenous vaccines, an increasing area for some animal species.

However, limitations exist with regard to the initial costs consisting of the equipment and continuous servicing. Furthermore, the technology is hardly acknowledged in official regulations, which still consider culture dependent methods as 'gold-standards'. MALDI-TOF MS should be part of a polyphasic approach combining traditional taxonomic and bacteriological methods together with nucleic acid analyses. Access to generated data, most likely via a public database, would not only establish a certain transparency it would also increase awareness and spread of the technology. Additionally, it would help to prevent misidentification by MALDI-TOF MS due to an insufficient number of well defined (reference) strains in the database, a serious problem especially in veterinary microbiology. Unfortunately, it would not prevent that a substantial amount of diagnostic data would still be available only at the laboratory in which they were generated but it would help to place MALDI TOF-MS on a comparable level to genetic data.

The establishment of robust, reproducible and reliable data will still be a major target for future investigations. Identification and discrimination of bacteria down to a subspecies level remains a challenge for numerous microorganisms and applications. Combining MALDI-TOF MS with bioinformatics to process spectra will create additional opportunities. Such systems should be freely accessible in the web to support dissemination of data. MALDI-TOF MS based clustering could be used as epidemiological method for evaluating bacterial disease in order to monitor the spread of an outbreak and to resolve where the infection started. This will be of special interest with regard to zoonotic pathogens, an area benefiting from the interest of human medicine.

Additionally, new approaches are needed to recruit the full potential of the technology in veterinary and food technology. Identification of bacteria directly from diseased animals or food products could be mentioned as an additional demand for the future. Implementing proteomics to identify contaminants and spoilage bacteria would increase the level of in-process controls. Establishing biomarkers for more bacteria would be helpful to improve species identification. The fast and substantial progress in sequencing technologies will help to correlate ion biomarker masses with protein masses, even though post-translational modifications need to be considered. Increasing efforts need to be done in order to extend the application of functional aspects offered by the technology. Determine antibiotic resistance combined with the identification of a certain bacteria would have great potential in clinical veterinary microbiology. Elucidating the mechanisms and beneficial effects of certain bacteria or a microbial community within the fermentation processes of the food industry bears numerous options for the technology. Again, new protocols have to be developed considering the broad range of matrices and bacteria to be targeted in veterinary and food microbiology.

As MALDI-TOF MS applied in veterinary and food microbiology has just started it is easy to predict that more indications and applications will come up in the near future, in order to comply with the huge demand offered by these areas.

References

Alispahic, M., Christensen, H., Hess, C., Razzazi-Fazeli, E., Bisgaard, M., and Hess, M. (2011). Identification of *Gallibacterium* species by matrix-assisted laser desorption/ionization time-of-flight mass spectrometry evaluated by multilocus sequence analysis. Int. J. Med. Microbiol. *301*, 513–522.

Alispahic, M., Christensen, H., Hess, C., Razzazi-Fazeli, E., Bisgaard, M., and Hess, M. (2012). MALDI-TOF mass spectrometry confirms clonal lineages of *Gallibacterium anatis* between chicken flocks. Vet. Microbiol. *160*, 269–273.

Alispahic, M., Christensen, H., Bisgaard, M., Hess, M., and Hess, C. (2014). MALDI-TOF mass spectrometry confirms difficulties in separating species of the *Avibacterium* genus. Avian Pathol. *43*, 258–263.

Alispahic, M., Hummel, K., Jandreski-Cvetkovic, D., Nobauer, K., Razzazi-Fazeli, E., Hess, M., and Hess, C. (2010). Species-specific identification and differentiation of *Arcobacter, Helicobacter* and *Campylobacter* by full-spectral matrix-associated laser desorption/ionization time of flight mass spectrometry analysis. J. Med. Microbiol. *59*, 295–301.

Balbutskaya, A., Sammra, O., Nagib, S., Hijazin, M., Alber, J., Lammler, C., Foster, G., Erhard, M., Wragg, P.N., Abdulmawjood, A., *et al.* (2014). Identification of *Arcanobacterium pluranimalium* by matrix-assisted laser desorption ionization-time of flight mass spectrometry and, as novel target, by sequencing pluranimaliumlysin encoding gene pla. Vet. Microbiol. *168*, 428–431.

Barbuddhe, S.B., Maier, T., Schwarz, G., Kostrzewa, M., Hof, H., Domann, E., Chakraborty, T., and Hain, T. (2008). Rapid identification and typing of *Listeria* species by matrix-assisted laser desorption ionization-time of flight mass spectrometry. Appl. Environ. Microbiol. *74*, 5402–5407.

Barreiro, J.R., Braga, P.A., Ferreira, C.R., Kostrzewa, M., Maier, T., Wegemann, B., Boettcher, V., Eberlin, M.N., and dos Santos, M.V. (2012). Nonculture-based identification of bacteria in milk by protein fingerprinting. Proteomics *12*, 2739–2745.

Barreiro, J.R., Ferreira, C.R., Sanvido, G.B., Kostrzewa, M., Maier, T., Wegemann, B., Bottcher, V., Eberlin, M.N., and dos Santos, M.V. (2010). Short communication: Identification of subclinical cow mastitis pathogens in milk by matrix-assisted laser desorption/ionization time-of-flight mass spectrometry. J. Dairy Sci. *93*, 5661–5667.

Benmechernene, Z., Fernandez-No, I., Quintela-Baluja, M., Böhme, K., Kihal, M., Calo-Mata, P., and Barros-Velazquez, J. (2014). Genomic and proteomic characterization of bacteriocin-producing *Leuconostoc mesenteroides* strains isolated from raw camel milk in two southwest Algerian arid zones. Biomed. Res. Int. *2014*, 853238.

Böhme, K., Fernandez No, I.C., Barros-Velazquez, J., Gallardo, J.M., Canas, B., and Calo-Mata, P. (2010a). Comparative analysis of protein extraction methods for the identification of seafood-borne pathogenic and spoilage bacteria by MALDI-TOF mass spectrometry. Analytical Methods *2*, 1941–1947.

Böhme, K., Fernandez-No, I.C., Barros-Velazquez, J., Gallardo, J.M., Calo-Mata, P., and Canas, B. (2010b). Species differentiation of seafood spoilage and pathogenic gram-negative bacteria by MALDI-TOF mass fingerprinting. J. Proteome Res. *9*, 3169–3183.

Böhme, K., Fernandez-No, I.C., Barros-Velazquez, J., Gallardo, J.M., Canas, B., and Calo-Mata, P. (2011). Rapid species identification of seafood spoilage and pathogenic Gram-positive bacteria by MALDI-TOF mass fingerprinting. Electrophoresis *32*, 2951–2965.

Böhme, K., Morandi, S., Cremonesi, P., Fernandez No, I.C., Barros-Velazquez, J., Castiglioni, B., Brasca, M., Canas, B., and Calo-Mata, P. (2012). Characterization of *Staphylococcus aureus* strains isolated from Italian dairy products by MALDI-TOF mass fingerprinting. Electrophoresis *33*, 2355–2364.

Böhme, K., Fernandez-No, I.C., Pazos, M., Gallardo, J.M., Barros-Velazquez, J., Canas, B., and Calo-Mata, P. (2013). Identification and classification of seafood-borne pathogenic and spoilage bacteria: 16S rRNA sequencing versus MALDI-TOF MS fingerprinting. Electrophoresis *34*, 877–887.

Bujnakova, D., Strakova, E., and Kmet, V. (2014). In vitro evaluation of the safety and probiotic properties of *Lactobacilli* isolated from chicken and calves. Anaerobe *29*, 118–127.

Bunesova, V., Vlkova, E., Rada, V., Rockova, S., Svobodova, I., Jebavy, L., and Kmet, V. (2012). *Bifidobacterium animalis* subsp. *lactis* strains isolated from dog faeces. Vet. Microbiol. *160*, 501–505.

Calderaro, A., Piccolo, G., Montecchini, S., Buttrini, M., Gorrini, C., Rossi, S., Arcangeletti, M.C., De Conto, F., Medici, M.C., and Chezzi, C. (2013). MALDI-TOF MS analysis of human and animal *Brachyspira* species and benefits of database extension. J. Proteomics *78*, 273–280.

Christensen, H., Bisgaard, M., Bojesen, A.M., Mutters, R., and Olsen, J.E. (2003). Genetic relationships among avian isolates classified as *Pasteurella haemolytica*, '*Actinobacillus salpingitidis*' or *Pasteurella anatis* with proposal of *Gallibacterium anatis* gen. nov., comb. nov. and description of additional genomospecies within *Gallibacterium* gen. nov. Int. J. Syst. Evol. Microbiol. *53*, 275–287.

Cools, P., Haelters, J., Lopes dos Santos Santiago, G., Claeys, G., Boelens, J., Leroux-Roels, I., Vaneechoutte, M., and Deschaght, P. (2013). *Edwardsiella tarda* sepsis in a live-stranded sperm whale (*Physeter macrocephalus*). Vet. Microbiol. *166*, 311–315.

Coyle, M.B., and Lipsky, B.A. (1990). Coryneform bacteria in infectious diseases: clinical and laboratory aspects. Clin. Microbiol. Rev. *3*, 227–246.

Dec, M., Urban-Chmiel, R., Gnat, S., Puchalski, A., and Wernicki, A. (2014). Identification of *Lactobacillus* strains of goose origin using MALDI-TOF mass spectrometry and 16S-23S rDNA intergenic spacer PCR analysis. Res. Microbiol. *165*, 190–201.

Decristophoris, P., Fasola, A., Benagli, C., Tonolla, M., and Petrini, O. (2011). Identification of *Staphylococcus intermedius* Group by MALDI-TOF MS. Syst. Appl. Microbiol. *34*, 45–51.

Dieckmann, R., Helmuth, R., Erhard, M., and Malorny, B. (2008). Rapid classification and identification of *Salmonellae* at the species and subspecies levels by whole-cell matrix-assisted laser desorption ionization-time of flight mass spectrometry. Appl. Environ. Microbiol. *74*, 7767–7778.

Dieckmann, R., and Malorny, B. (2011). Rapid screening of epidemiologically important *Salmonella* enterica subsp. *enterica* serovars by whole-cell matrix-assisted laser desorption ionization-time of flight mass spectrometry. Appl. Environ. Microbiol. *77*, 4136–4146.

Djelouadji, Z., Roux, V., Raoult, D., Kodjo, A., and Drancourt, M. (2012). Rapid MALDI-TOF mass spectrometry identification of *Leptospira* organisms. Vet. Microbiol. *158*, 142–146.

Duskova, M., Sedo, O., Ksicova, K., Zdrahal, Z., and Karpiskova, R. (2012). Identification of *Lactobacilli* isolated from food by genotypic methods and MALDI-TOF MS. Int. J. Food Microbiol. *159*, 107–114.

Fagerquist, C.K., Miller, W.G., Harden, L.A., Bates, A.H., Vensel, W.H., Wang, G., and Mandrell, R.E. (2005). Genomic and proteomic identification of a DNA-binding protein used in the 'fingerprinting' of *Campylobacter* species and strains by MALDI-TOF-MS protein biomarker analysis. Anal. Chem. *77*, 4897–4907.

Fagerquist, C.K., Bates, A.H., Heath, S., King, B.C., Garbus, B.R., Harden, L.A., and Miller, W.G. (2006). Sub-speciating *Campylobacter jejuni* by proteomic analysis of its protein biomarkers and their post-translational modifications. J. Proteome Res. *5*, 2527–2538.

Fagerquist, C.K. (2007). Amino acid sequence determination of protein biomarkers of *Campylobacter upsaliensis* and *C. helveticus* by 'composite' sequence proteomic analysis. J. Proteome Res. *6*, 2539–2549.

Fagerquist, C.K., Yee, E., and Miller, W.G. (2007). Composite sequence proteomic analysis of protein biomarkers of *Campylobacter coli*, *C. lari* and *C. concisus* for bacterial identification. Analyst *132*, 1010–1023.

Favaro, L., Basaglia, M., Casella, S., Hue, I., Dousset, X., Gombossy de Melo Franko, B.D., and Todorov, S.D. (2014). Bacteriocinogenic potential and safety evaluation of non-starter *Enterococcus faecium* strains isolates from home made white brine cheese. Food Microbiol. *38*, 228–239.

Fernandez-No, I.C., Böhme, K., Gallardo, J.M., Barros-Velazquez, J., Canas, B., and Calo-Mata, P. (2010). Differential characterization of biogenic amine-producing bacteria involved in food poisoning using MALDI-TOF mass fingerprinting. Electrophoresis *31*, 1116–1127.

Fernandez-No, I.C., Böhme, K., Calo-Mata, P., and Barros-Velazquez, J. (2011). Characterisation of histamine-producing bacteria from farmed blackspot seabream (*Pagellus bogaraveo*) and turbot (*Psetta maxima*). Int. J. Food Microbiol. *151*, 182–189.

Fernandez-No, I.C., Böhme, K., Diaz-Bao, M., Cepeda, A., Barros-Velazquez, J., and Calo-Mata, P. (2013). Characterisation and profiling of *Bacillus subtilis*, *Bacillus cereus* and *Bacillus licheniformis* by MALDI-TOF mass fingerprinting. Food Microbiol. 33, 235–242.

Goncalves, J.L., Tomazi, T., Barreiro, J.R., Braga, P.A., Ferreira, C.R., Araujo Junior, J.P., Eberlin, M.N., and dos Santos, M.V. (2014). Identification of *Corynebacterium* spp. isolated from bovine intramammary infections by matrix-assisted laser desorption ionization-time of flight mass spectrometry. Vet. Microbiol. *173*, 147–151.

Goto, K., Yamamoto, M., Asahara, M., Tamura, T., Matsumura, M., Hayashimoto, N., and Makimura, K. (2012). Rapid identification of *Mycoplasma pulmonis* isolated from laboratory mice and rats using matrix-assisted laser desorption ionization time-of-flight mass spectrometry. J. Vet. Med. Sci. *74*, 1083–1086.

Grosse-Herrenthey, A., Maier, T., Gessler, F., Schaumann, R., Bohnel, H., Kostrzewa, M., and Kruger, M. (2008). Challenging the problem of clostridial identification with matrix-assisted laser desorption and ionization-time-of-flight mass spectrometry (MALDI-TOF MS). Anaerobe *14*, 242–249.

Hess, C., Enichlmayr, H., Jandreski-Cvetkovic, D., Liebhart, D., Bilic, I., and Hess, M. (2013). *Riemerella anatipestifer* outbreaks in commercial goose flocks and identification of isolates by MALDI-TOF mass spectrometry. Avian Pathol. *42*, 151–156.

Hijazin, M., Hassan, A.A., Alber, J., Lammler, C., Timke, M., Kostrzewa, M., Prenger-Berninghoff, E., and Zschock, M. (2012a). Evaluation of matrix-assisted laser desorption ionization-time of flight mass spectrometry (MALDI-TOF MS) for species identification of bacteria of genera *Arcanobacterium* and *Trueperella*. Vet. Microbiol. *157*, 243–245.

Hijazin, M., Metzner, M., Erhard, M., Nagib, S., Alber, J., Lammler, C., Hassan, A.A., Prenger-Berninghoff, E., and Zschock, M. (2012b). First description of *Trueperella* (*Arcanobacterium*) *bernardiae* of animal origin. Vet. Microbiol. *159*, 515–518.

Horka, M., Karasek, P., Salplachta, J., Ruzicka, F., Vykydalova, M., Kubesova, A., Drab, V., Roth, M., and Slais, K. (2013). Capillary isoelectric focusing of probiotic bacteria from cow's milk in tapered fused silica capillary with off-line matrix-assisted laser desorption/ionization time-of-flight mass spectrometry identification. Anal. Chim. Acta *788*, 193–199.

Huber, H., Ziegler, D., Pfluger, V., Vogel, G., Zweifel, C., and Stephan, R. (2011). Prevalence and characteristics of methicillin-resistant coagulase-negative staphylococci from livestock, chicken carcasses, bulk tank milk, minced meat, and contact persons. BMC Vet. Res. *7*, 6–6148–7–6.

Jadhav, S., Sevior, D., Bhave, M., and Palombo, E.A. (2014). Detection of *Listeria monocytogenes* from selective enrichment broth using MALDI-TOF Mass Spectrometry. J. Proteomics *97*, 100–106.

Jay-Russell, M.T., Bates, A., Harden, L., Miller, W.G., and Mandrell, R.E. (2012). Isolation of *Campylobacter* from feral swine (Sus scrofa) on the ranch associated with the 2006 *Escherichia coli* O157:H7 spinach outbreak investigation in California. Zoonoses Public. Health. *59,* 314–319.

Kern, C.C., Usbeck, J.C., Vogel, R.F., and Behr, J. (2013). Optimization of Matrix-Assisted-Laser-Desorptio n-Ionization-Time-Of-Flight Mass Spectrometry for the identification of bacterial contaminants in beverages. J. Microbiol. Methods *93,* 185–191.

Kern, C.C., Vogel, R.F., and Behr, J. (2014a). Differentiation of *Lactobacillus brevis* strains using Matri x-Assisted-Laser-Desorption-Ionization-Time-of-Flight Mass Spectrometry with respect to their beer spoilage potential. Food Microbiol. *40,* 18–24.

Kern, C.C., Vogel, R.F., and Behr, J. (2014b). Identification and differentiation of brewery isolates of *Pectinatus* sp. by Matrix-Assisted-Laser Desorption-Ionization Time-Of-Flight Mass Spectrometry (MALDI-TOF MS). European Food Research and Technology *238,* 875–880.

Klein, D., Alispahic, M., Sofka, D., Iwersen, M., Drillich, M., and Hilbert, F. (2013). Prevalence and risk factors for shedding of thermophilic *Campylobacter* in calves with and without diarrhea in Austrian dairy herds. J. Dairy Sci. *96,* 1203–1210.

Kolinska, R., Drevinek, M., Jakubu, V., and Zemlickova, H. (2008). Species identification of *Campylobacter jejuni* ssp. *jejuni* and *C. coli* by matrix-assisted laser desorption/ionization time-of-flight mass spectrometry and PCR. Folia Microbiol. (Praha) *53,* 403–409.

Kuda, T., Izawa, Y., Yoshida, S., Koyanagi, T., Takahashi, H., and Kimura, B. (2014). Rapid identification of *Tetragenococcus halophilus* and *Tetragenococcus muriaticus,* important species in the production of salted and fermented foods, by matrix-assisted laser desorption ionization-time of flight mass spectrometry (MALDI-TOF MS). Food Control *35,* 419–425.

Kuhnert, P., Bisgaard, M., Korczak, B.M., Schwendener, S., Christensen, H., and Frey, J. (2012). Identification of animal *Pasteurellaceae* by MALDI-TOF mass spectrometry. J. Microbiol. Methods *89,* 1–7.

Mandrell, R.E., Harden, L.A., Bates, A., Miller, W.G., Haddon, W.F., and Fagerquist, C.K. (2005). Speciation of *Campylobacter coli, C. jejuni, C. helveticus, C. lari, C. sputorum,* and *C. upsaliensis* by matrix-assisted laser desorption ionization-time of flight mass spectrometry. Appl. Environ. Microbiol. *71,* 6292–6307.

Mandrell, R.E., and Wachtel, M.R. (1999). Novel detection techniques for human pathogens that contaminate poultry. Curr. Opin. Biotechnol. *10,* 273–278.

Masarikova, M., Mrackova, M., and Sedlinska, M. (2014). Application of Matrix-Assisted Laser Desorption Ionization Time-of-Flight Mass Spectrometry in Identiifcation of Stallion Semen Bacterial Contamination. J. Equine Vet. Sci. *34,* 833–836.

Mather, C.A., Rivera, S.F., and Butler-Wu, S.M. (2014). Comparison of the Bruker Biotyper and Vitek MS matrix-assisted laser desorption ionization-time of flight mass spectrometry systems for identification of mycobacteria using simplified protein extraction protocols. J. Clin. Microbiol. *52,* 130–138.

Mazzeo, M.F., Sorrentino, A., Gaita, M., Cacace, G., Di Stasio, M., Facchiano, A., Comi, G., Malorni, A., and Siciliano, R.A. (2006). Matrix-assisted laser desorption ionization-time of flight mass spectrometry for the discrimination of food-borne microorganisms. Appl. Environ. Microbiol. 72, 1180–1189.

Metzner, M., Erhard, M., Sammra, O., Nagib, S., Hijazin, M., Alber, J., Lammler, C., Abdulmawjood, A., Prenger-Berninghoff, E., Zschock, M., *et al.* (2013). *Trueperella abortisuis,* an emerging pathogen isolated from pigs in Germany. Berl. Munch. Tierarztl. Wochenschr. *126,* 423–426.

Paudel, S., Alispahic, M., Liebhart, D., Hess, M., and Hess, C. (2013). Assessing pathogenicity of *Gallibacterium anatis* in a natural infection model: the respiratory and reproductive tracts of chickens are targets for bacterial colonization. Avian Pathol. *42,* 527–535.

Pereyre, S., Tardy, F., Renaudin, H., Cauvin, E., Del Pra Netto Machado, L., Tricot, A., Benoit, F., Treilles, M., and Bebear, C. (2013). Identification and subtyping of clinically relevant human and ruminant *Mycoplasmas* by use of matrix-assisted laser desorption ionization-time of flight mass spectrometry. J. Clin. Microbiol. *51,* 3314–3323.

Petternel, C., Galler, H., Zarfel, G., Luxner, J., Haas, D., Grisold, A.J., Reinthaler, F.F., and Feierl, G. (2014). Isolation and characterization of multidrug-resistant bacteria from minced meat in Austria. Food Microbiol. *44,* 41–46.

Philipp, H.C., Taras, D., Liman, M., Grosse-Herrenthey, A., Rönchen, S., and Behr, P. (2013). Molecular typing of *Riemerella anatipestifer* serotype 14, an emerging pathogen for ducks. Archiv Für Geflügelkunde *77,* 218–225.

Prohaska, S., Pfluger, V., Ziegler, D., Scherrer, S., Frei, D., Lehmann, A., Wittenbrink, M.M., and Huber, H. (2014). MALDI-TOF MS for identification of porcine *Brachyspira* species. Lett. Appl. Microbiol. *58,* 292–298.

Quintela-Baluja, M., Böhme, K., Fernandez-No, I.C., Morandi, S., Alnakip, M.E., Caamano-Antelo, S., Barros-Velazquez, J., and Calo-Mata, P. (2013). Characterization of different food-isolated *Enterococcus* strains by MALDI-TOF mass fingerprinting. Electrophoresis *34*, 2240–2250.

Rubbenstroth, D., Hotzel, H., Knobloch, J., Teske, L., Rautenschlein, S., and Ryll, M. (2011). Isolation and characterization of atypical *Riemerella columbina* strains from pigeons and their differentiation from *Riemerella anatipestifer*. Vet. Microbiol. *147*, 103–112.

Rubbenstroth, D., Ryll, M., Hotzel, H., Christensen, H., Knobloch, J.K., Rautenschlein, S., and Bisgaard, M. (2013a). Description of *Riemerella columbipharyngis* sp. nov., isolated from the pharynx of healthy domestic pigeons (*Columba livia f. domestica*), and emended descriptions of the genus *Riemerella, Riemerella anatipestifer* and *Riemerella columbina*. Int. J. Syst. Evol. Microbiol. *63*, 280–287.

Rubbenstroth, D., Ryll, M., Knobloch, J.K., Kohler, B., and Rautenschlein, S. (2013b). Evaluation of different diagnostic tools for the detection and identification of *Riemerella anatipestifer*. Avian Pathol. *42*, 17–26.

Rzewuska, M., Witkowski, L., Cisek, A.A., Stefanska, I., Chrobak, D., Stefaniuk, E., Kizerwetter-Swida, M., and Takai, S. (2014). Characterization of *Rhodococcus equi* isolates from submaxillary lymph nodes of wild boars (*Sus scrofa*), red deer (*Cervus elaphus*) and roe deer (*Capreolus capreolus*). Vet. Microbiol. *172*, 272–278.

Sammra, O., Balbutskaya, A., Zhang, S., Hijazin, M., Nagib, S., Lammler, C., Abdulmawjood, A., Prenger-Berninghoff, E., Kostrzewa, M., and Timke, M. (2013). Further characteristics of *Arcanobacterium canis*, a novel species of genus *Arcanobacterium*. Vet. Microbiol. *167*, 619–622.

Sammra, O., Balbutskaya, A., Nagib, S., Alber, J., Lammler, C., Abdulmawjood, A., Timke, M., Kostrzewa, M., and Prenger-Berninghoff, E. (2014). Properties of an *Arcanobacterium haemolyticum* strain isolated from a donkey. Berl. Munch. Tierarztl. Wochenschr. *127*, 56–60.

Schäfer, M.O., Genersch, E., Funfhaus, A., Poppinga, L., Formella, N., Bettin, B., and Karger, A. (2014). Rapid identification of differentially virulent genotypes of *Paenibacillus larvae*, the causative organism of American foulbrood of honey bees, by whole cell MALDI-TOF mass spectrometry. Vet. Microbiol. *170*, 291–297.

Seibold, E., Maier, T., Kostrzewa, M., Zeman, E., and Splettstoesser, W. (2010). Identification of *Francisella tularensis* by whole-cell matrix-assisted laser desorption ionization-time of flight mass spectrometry: fast, reliable, robust, and cost-effective differentiation on species and subspecies levels. J. Clin. Microbiol. *48*, 1061–1069.

Spitaels, F., Wieme, A.D., Janssens, M., Aerts, M., Daniel, H.M., Van Landschoot, A., De Vuyst, L., and Vandamme, P. (2014). The microbial diversity of traditional spontaneously fermented lambic beer. PLoS One *9*, e95384.

Sun, L., Teramoto, K., Sato, H., Torimura, M., Tao, H., and Shintani, T. (2006). Characterization of ribosomal proteins as biomarkers for matrix-assisted laser desorption/ionization mass spectral identification of *Lactobacillus plantarum*. Rapid Commun. Mass Spectrom. *20*, 3789–3798.

Teramoto, K., Sato, H., Sun, L., Torimura, M., and Tao, H. (2007). A simple intact protein analysis by MALDI-MS for characterization of ribosomal proteins of two genome-sequenced lactic acid bacteria and verification of their amino acid sequences. J. Proteome Res. *6*, 3899–3907.

Tomazi, T., Goncalves, J.L., Barreiro, J.R., de Campos Braga, P.A., Prada e Silva, L.F., Eberlin, M.N., and dos Santos, M.V. (2014). Identification of coagulase-negative staphylococci from bovine intramammary infection by matrix-assisted laser desorption ionization-time of flight mass spectrometry. J. Clin. Microbiol. *52*, 1658–1663.

Vithanage, N.R., Yeager, T.R., Jadhav, S.R., Palombo, E.A., and Datta, N. (2014). Comparison of identification systems for psychrotrophic bacteria isolated from raw bovine milk. Int. J. Food Microbiol. *189*, 26–38.

Wenning, M., Breitenwieser, F., Konrad, R., Huber, I., Busch, U., and Scherer, S. (2014). Identification and differentiation of food-related bacteria: A comparison of FTIR spectroscopy and MALDI-TOF mass spectrometry. J. Microbiol. Methods *103*, 44–52.

Werner, G., Fleige, C., Fessler, A.T., Timke, M., Kostrzewa, M., Zischka, M., Peters, T., Kaspar, H., and Schwarz, S. (2012). Improved identification including MALDI-TOF mass spectrometry analysis of group D streptococci from bovine mastitis and subsequent molecular characterization of corresponding *Enterococcus faecalis* and *Enterococcus faecium* isolates. Vet. Microbiol. *160*, 162–169.

Wieme, A.D., Spitaels, F., Aerts, M., De Bruyne, K., Van Landschoot, A., and Vandamme, P. (2014a). Effects of growth medium on matrix-assisted laser desorption-ionization time of flight mass spectra: a case study of acetic acid bacteria. Appl. Environ. Microbiol. *80*, 1528–1538.

Wieme, A.D., Spitaels, F., Aerts, M., De Bruyne, K., Van Landschoot, A., and Vandamme, P. (2014b). Identification of beer-spoilage bacteria using matrix-assisted laser desorption/ionization time-of-flight mass spectrometry. Int. J. Food Microbiol. *185*, 41–50.

Winkler, M.A., Uher, J., and Cepa, S. (1999). Direct analysis and identification of *Helicobacter* and *Campylobacter* species by MALDI-TOF mass spectrometry. Anal. Chem. *71*, 3416–3419.

Wragg, P., Randall, L., and Whatmore, A.M. (2014). Comparison of Biolog GEN III MicroStation semi-automated bacterial identification system with matrix-assisted laser desorption ionization-time of flight mass spectrometry and 16S ribosomal RNA gene sequencing for the identification of bacteria of veterinary interest. J. Microbiol. Methods *105*, 16–21.

Yassin, A.F., Hupfer, H., Siering, C., and Schumann, P. (2011). Comparative chemotaxonomic and phylogenetic studies on the genus *Arcanobacterium* Collins *et al.* 1982 emend. Lehnen *et al.* 2006: proposal for *Trueperella* gen. nov. and emended description of the genus *Arcanobacterium*. Int. J. Syst. Evol. Microbiol. *61*, 1265–1274.

MALDI-TOF MS for Environmental Analysis, Microbiome Research and as a Tool for Biological Resource Centres

7

Markus Kostrzewa, Chantal Bizet and Dominique Clermont

Abstract

The simplicity of sample preparation, speed of analysis, high-throughput capabilities and low analysis costs per sample have made MALDI-TOF MS profiling not only a successful first-line diagnostic method but also a favoured tool for research and analysis of environmental strains and microbial communities. Its differentiation potential, which is frequently higher than that of 16S rDNA sequence analysis, enables detection of new species and subspecies and the dereplication of large numbers of isolated microbes. Further, tracking of environmental strains seems to be feasible at least for particular species. Thereby, MALDI-TOF MS besides DNA-based and biochemical methods becomes a further basic tool for research in environmental microbiology and phylogeny. These characteristics also have made MALDI-TOF MS fingerprinting a first-line method for species identification and confirmation in strain collections. Together with other complementary methods it is already established for one factor in the multidimensional approach for description of new species.

Introduction

Besides routine diagnostic or diagnostic-like purposes in clinic, veterinary, pharma and food microbiology, MALDI-TOF MS fingerprinting also has evolved as a powerful tool for analysis of organisms in environment and in the field of microbial community research. The throughput capabilities, accuracy and low running costs of a MALDI-TOF MS system enable analyses in a scale which were just not possible until recently. Thereby, an approach called 'culturomics' to investigate microbial diversity has been proposed, based on conventional culture using a variety of culture conditions, MALDI-TOF MS profiling, and DNA sequence analysis for those organisms, which could not be identified by MALDI-TOF MS (Lagier *et al.*, 2012). This approach has been shown to be a powerful alternative or complement to culture-free methods for microbiome analysis. Also in a smaller scale MALDI-TOF MS has been shown to be a valuable tool for research in particular environments, for water-, soil- and air-derived organisms. Here, especially the capability of MALDI-TOF MS to identify many rare microbes and to cluster even recently non-identified microorganisms in 'MALDI groups', which indicate a close relationship, are advantageous. In addition, tracking of environmental contamination routes is possible at least for species with sufficient intra-species diversity of profile spectra.

Biological Resource Centers also have found that MALDI-TOF MS is a very strong tool for their strain control. As the method is getting one tool in multiphasic description of new species and a respective MALDI-TOF MS profile spectrum can be stored as a reference for each newly defined species and strain, it is straight forward to introduce MALDI-TOF MS for rapid and accurate species confirmation. Detailed spectra comparison may even enable strain confirmation to a certain extend.

MALDI-TOF MS for microbiome analysis: microbial culturomics

Recently, a novel cultivation-based approach has been proposed for the investigation of large complex microbial communities. This approach uses a plurality of different cultivation conditions to obtain as many as possible microbial colonies from a particular environment. Subsequently, an identification scheme based on MALDI-TOF MS fingerprinting and molecular analysis, i.e. 16s rRNA gene sequencing, is applied to speciate as many microorganisms as possible and cluster them based on the analysis results. Key for the applicability of this approach is the high number or organisms, which can be identified quickly and very cost-effective by MALDI-TOF MS. Gaps in the database can be closed by the utilization of gene sequencing. Therefore, at affordable costs and with reasonable efforts a very high number of isolates can be analysed with high accuracy. Initially, this new approach has been applied to the complex microbiome of the gut. The first study reporting this method was applied to human gut microbiota (Lagier *et al.*, 2012). In this study, stools from two lean Africans and one obese European using 212 different culture conditions were investigated. The colonies grown under the variety of conditions were investigated using MALDI-TOF MS and 16S rRNA amplification sequencing. A total of 32,500 colonies obtained by the different cultures yielded 340 different bacterial species from seven phyla and 117 genera, including two species from rare phyla (*Deinococcus-Thermus* and Synergistetes), five fungi and a giant virus (Senegalvirus). Interestingly, the part of the microbiome identified by this new strategy included 174 species, which never had been described in the human gut before. On the other hand, metagenomic analysis of the same samples revealed 698 phylotypes, including 282 known species. Only 51 of these species overlapped with the microbiome identified by the culturomics approach. The authors concluded that obviously microbial culturomics complements metagenomics by overcoming the depth bias inherent in metagenomics. In a proof-of-concept study for investigation of atypical stool samples, the gut microbiome of a patient with resistant tuberculosis has been investigated (Dubourg *et al.*, 2013). Culturomics here was found to be an even superior technique compared to metagenomics. Pfleiderer and colleagues studied the gut microbiome of an anorexia nervosa patient and compared culturomics with a metagenomics approach applying pyrosequencing of 16S rDNA targeting the V6 region (Pfleiderer *et al.*, 2013). In total, 88 culture conditions generating 12,700 different colonies were investigated. Thereby, the authors have identified 133 bacterial species, with 19 bacterial species never isolated from the human gut before, including 11 new bacterial species, for which the genome has been sequenced. Interestingly, the overlap between pyrosequencing/metagenomics and culturomics was only 17%. The authors concluded that their study allowed extending more significantly the repertoire of the human gut microbiome than bacterial species validated by the rest of the world during the two years before. As most of the bacteria detected by pyrosequencing/metagenomics only did belong to strictly anaerobic species, an important complementarity of both methods has been demonstrated. In another investigation, culturomics was applied to the gut microbiome of four patients under treatment with large-scale antibiotics (Dubourg *et al.*, 2014). Again, pyrosequencing of the V6 region was applied in parallel and gut richness was estimated by bacterial counting after microscopic observation. The culturomics approach tested 77 culture conditions, resulting in 32,000 different colonies that were investigated. Overall, 190 bacterial species were identified, with nine species that had never been isolated from the human gut before. The study detected seven species newly described in humans and eight completely new species. A dramatic reduction in diversity was observed for two of the four stool samples for which antibiotic treatment was prolonged and uninterrupted. Gouba *et al.* applied culturomics to the eukaryotic gut microbiome, which is much less explored than the bacterial gut microbiome (Gouba *et al.*, 2014). The authors analysed this part of the microbiome in seven individuals living in four tropical countries. In parallel to culture they utilized PCR-sequencing. Thereby, a total of 41 microeukaryotes including 38 different fungal species and three protists were detected. While four fungal species were only detected by the culturomics technique, even 27 where uniquely found by the PCR-based analyses. The results again were depicting the complementary of both approaches. Culturomics was also applied as a part of animal gut microbiome analysis. Bittar and colleagues investigated the gorilla gut microbiome as a source for human pathogens (Bittar *et al.*, 2014). In total, 48 faecal samples obtained from 21 *Gorilla gorilla gorilla* individuals were screened for human bacterial pathogens using molecular techniques. For one index gorilla, using specific media supplemented by plants, 12,800 colonies were obtained. Using this complementary study design of culturomics and metagenomics, 147 different bacterial species were identified, including many opportunistic pathogens and five new species. In a study to investigate the bacterial colonization of intensive care patients and the effect of chlorhexidine (CHG) daily bathing on it (Cassir *et al.*, 2015), twenty ICU patients were included, 10 of them during an interventional period with CHG daily bathing, ten for a control period without CHG bathing. At day 7 of hospitalization, eight skin swab samples from nares, axillary vaults, inguinal creases, manubrium and back, were taken from each patient. The study obtained 5,000 colonies that yielded 61 bacterial species (9.15 ± 3.7 per patient), including 15 that had never been cultured from non-pathological human skin before. Three of these species had never been cultured from human samples before. While in the control group a higher risk of colonization with Gram-negative bacteria was found, in the CHG group more patients were colonized by sporulating bacteria and showed a reduced skin bacterial richness and lower microbial diversity. Further studies applying culturomics to non-gut microbiomes have to be expected, which will contribute to our knowledge about microbiome diversity and functions.

MALDI-TOF MS profiling in environmental analysis

MALDI-TOF MS has also been identified as a powerful tool in environmental microbial analysis. Again, the potential for high-throughput analyses is one key factor, further the ability to build own libraries of interest, e.g. for the microbes of a particular environment. The capability to compare even strains with yet unassigned species name with other known and unknown strains and isolates by similarity dendrogram calculation or other statistical analyses makes the technology a valuable phylogenetic analysis tool.

Water is one of the habitats where MALDI-TOF MS microbial profiling has successfully been applied. Here, in particular the genus *Vibrio*, as several of its species are potential human pathogens and because of its widespread occurrence, has been investigated. Hazen and co-authors found MALDI-TOF MS suitable for the distinction of *V. parahaemolyticus* from other *Vibrio* spp. (Hazen *et al.*, 2009). The MALDI-TOF MS spectra of *V. parahaemolyticus* strains were found to be distinct from those of the other *Vibrio* species examined in the study including the closely related species *V. alginolyticus*, *V. harveyi*, and *V. campbellii*. Furthermore, the authors found a significant spectra variability in the *V. parahaemolyticus* strains investigated, i.e. between spectra obtained from strains isolated from different geographical locations and at different times. Dieckmann and coworkers applied MALDI-TOF MS to the identification and characterization of a variety of *Vibrio* spp. (Dieckmann *et al.*, 2010). They found that the MALDI-TOF MS-based method did work as good as an *rpoB* sequence-based approach for *Vibrio* identification. In this study, MALDI-TOF MS and *rpoB* sequences resulted in very similar phylogenetic trees pointing to a very good congruence of both methods. In another study, MALDI-TOF MS was proposed as part of a polyphasic approach for *Vibrio* spp. characterization (Oberbeckmann *et al.*, 2011). Here, the mass spectrometry method was used as a primary screen to classify isolates, 16S rRNA gene and *rpoB* gene sequencing to finally identify the species. Potential *V. parahaemolyticus* isolates were screened for regulatory or virulence-related genes (*toxR, tlh, tdh, trh*). To investigate genomic diversity, repetitive-sequence-based PCRs were applied. Higher *Vibrio* abundances and infections are predicted for northern Europe due to climate changes. Environmental surveillance programs to evaluate this risk are necessary, therefore a German research consortium constructed a dedicated database for *Vibrio* spp. identification (Erler *et al.*, 2015). The consortium proposed that MALDI-TOF MS profiling could be used for the fast and reliable species classification of environmental isolates. Because the commercially available reference database did not contain sufficient *Vibrio* spectra, the VibrioBase database was established containing mass profile references from 997 mainly environmental strains. MALDI-TOF MS clusters in this study were assigned based on the species classification obtained by analysis of partial *rpoB* sequences. The affiliation of strains to species-specific clusters was consistent in 97% of all cases using both approaches. Importantly, the database created by the consortium (VibrioBase) was made freely accessible after the study. MALDI-TOF MS profiling was successfully applied to identification and even strain–group clustering of *Vibrio parahaemolyticus* from different sources (Malainine *et al.*, 2013) and reliable strain grouping also has been shown by Eddabra and co-workers (Eddabra *et al.*, 2012).

Ballast water, as a special water habitat, was investigated by Emami *et al.* (2012). In their study, marine bacteria and bacteria obtained from artificially created ballast water were identified by both MALDI-TOF MS and 16S rRNA gene sequencing. MALDI-TOF MS revealed the same identification results at the genus-level for 36 isolates. Donohue and colleagues reported the identification of *Aeromonas* spp. from drinking water, a genus which contains a number of human pathogenic species, with good concordance to phenotypic tests (Donohue *et al.*, 2007). A library of reference spectra of 40 strains from 17 different species was established for this purpose and challenged with 52 isolates in two blind studies, one with phenotypically typical and the other with atypical isolates, respectively. While all typical isolates were identified congruently, results from MALDI-TOF were concordant for

18 of 27 atypical isolates. The benefit of a MALDI-TOF MS database even containing references for diagnostically rare, water-born microorganisms was described for a 64-year-old male patient, returning from Australia to Germany (Hundenborn *et al.*, 2013). A wound of this man by MALDI-TOF analysis was found to be infected with two marine bacteria, *Vibrio harvei* which has never been reported for a wound infection before and *Photobacterium damsellae* which has been found in wounds rarely, respectively. Further studies have described the utilization of MALDI-TOF MS profiling for identification and classification of water-born organisms, partially in the context of multiphasic approaches (Martinez-Murcia *et al.*, 2008, Ali *et al.*, 2009, Ng *et al.*, 2014, Levican *et al.*, 2015).

Also microorganisms living in soil have been analysed by MALDI-TOF MS profiling, successfully. Behrendt and co-workers did investigate heterotrophic nitrifying bacteria with respiratory ammonification and denitrification activity (Behrendt *et al.*, 2010). Using MALDI-TOF MS profiling as part of a multiphasic approach they could describe two related new *Paenibacillus* species, one from a spacecraft assembly clean room, the other from fen soil. In another study, MALDI-TOF MS supplemented by 16S rRNA gene sequencing was applied to identify microorganisms for bioremediation, i.e. of bacteria that can metabolize biphenyl isolated from contaminated soil (Uhlik *et al.*, 2011). Even non-identified microorganisms could be analysed for their relatedness by dendrogram calculation based on their profile mass spectra. In some cases, MALDI-TOF MS was found to be more informative than sequence analysis where 16S rDNA sequencing could only identify a group, but not distinguish the contained species.

MALDI-TOF MS has further been applied to the identification and characterization of airborne microorganisms. Fox and colleagues speciated micrococci and staphylococci collected from schoolroom air by MALDI-TOF MS profile analysis (Fox *et al.*, 2010). They found that micrococci were much more frequent in this environment than previously reported for clinical samples. They concluded that this could lead to a misidentification of staphylococci without an appropriate species identification tool, e.g. MALDI-TOF MS. Similarly, MALDI-TOF MS has been utilized in studies for the characterization of the air from underground subway stations (Dybwad *et al.*, 2012) and bioaerosols (Madsen *et al.*, 2015).

Microbes living on or in community with plants have been investigated by MALDI-TOF MS fingerprinting. Behrendt and co-workers performed a taxonomic study of *Pseudomonas cedrina* strains isolated from the phyllosphere of grasses. They could differentiate a group of strains clearly from the type strain of *P. cedrina* (Behrendt *et al.*, 2009). The differences found between the fingerprint spectra of these isolates and of the type strain of *P. cedrina* were more significant than those detectable between the type strain and other, phylogenetically closely related species of the genus *Pseudomonas*. Together with results obtained from other analytical methods, this leads to the split of the species into two subspecies, i.e. *Pseudomonas cedrina* subsp. *cedrina* and *Pseudomonas cedrina* subsp. *fulgida*. Sauer *et al.* (2008) demonstrated the applicability of MALDI-TOF MS to identify and differentiate *Erwinia* species, a group of bacteria which includes a number of important plant pathogens. In a later study, the technique was used as a gold standard to validate PCR primers for differentiation of *Erwinia* species (Wensing *et al.*, 2012). Similarily, another plant pathogen, *Pantoea stewartii*, was demonstrated to be successfully identified and differentiated by MALDI-TOF MS fingerprinting (Wensing *et al.*, 2010; Gehring *et al.*, 2014). In a more comprehensive approach, Hausdorf and coworkers investigated spinach and wash water samples taken of the complete process line of a spinach-washing plant. 451 bacteria colonies grown on plate-count, *Arcobacter* selective, marine and blood agar were evaluated by MALDI-TOF MS fingerprinting, 16S rRNA gene sequencing and phylogenetic analysis (Hausdorf *et al.*, 2013). This approach revealed that a broad variety of microbes were present on the plants and in the wash water, including many isolated species from genera, which contain pathogenic or opportunistic pathogenic bacteria. A further interesting finding was that bacterial diversity on the spinach surface increased after the first washing step indicating a process-borne contamination of the spinach.

An area of scientific research where MALDI-TOF MS with its beneficial characteristics in speed and costs showed to be of particular advantage is so called 'dereplication' of microorganisms. Dereplication means the grouping of larger sets of isolates of unknown organisms without detailed taxonomic (e.g. species) identification to assess the diversity of the microorganisms found in a particular habitat. This is of special importance for microorganisms from habitats where many microorganisms are from yet not described phylogenetic origin. The approach can save a lot of time, work and money in large microbial community studies. As an example, microorganisms from different soil or water environments are not well investigated today and include a large number of isolates that have not been systematically described yet. Here, MALDI-TOF MS profiling can be used to rapidly generate raw data of larger communities at low operating cost. Then, the profiles can be compared and grouped according to criteria (e.g. matching thresholds, profile similarity) determined and defined earlier. The identification of index organisms can even be used to position the groups in the described taxonomic system. One early study, which investigated the suitability of MALDI-TOF MS for this purpose, analysed 456 bacterial isolates obtained from marine sponges. These sponges lived in a depth of about 300 m in the North Sea near to the Norwegian coast (Dieckmann *et al.*, 2005). Clustering into 11 groups (i.e. genera) could be obtained and was confirmed by 16S rDNA sequencing. For one group even a further subgrouping was achieved. In a further study, rep-PCR (repetitive element sequence based polymerase chain reaction) was used as benchmark technology, an established method for dereplication (Ghyselinck *et al.*, 2011). For 204 of 249 (82%) unidentified bacterial isolates, which had been retrieved from the rhizosphere of potato plants, the taxonomic resolution of both techniques was comparable, while for the remaining 45 isolates (18%) one of both revealed a higher resolution. 16S rDNA sequencing was used as further control method and showed that MALDI-TOF MS was the more reproducible method. Its taxonomic resolution was found to be between species and strain level. With its high-throughput and automation capabilities at low cost, MALDI-TOF MS is the more promising dereplication technology according to the authors. Further studies, in which the successful utilization of MALDI-TOF MS profiling is used for diversity investigation of microorganisms, have been described for microalgae (Emami *et al.*, 2015), halophilic prokaryotes (Munoz *et al.*, 2011) and bacteria from a cave environment (Zhang *et al.*, 2015).

Because of its capabilities to analyse biodiversity MALDI-TOF MS has been applied to source tracking of microorganisms, e.g. to find the source of a pollution. Thevenon and coworkers have described to use the technique in combination with specific PCR for the characterization of faecal indicator bacteria (Thevenon *et al.*, 2012). They could demonstrate for Lake Geneva that human faecal bacteria were highly increased in the sediments contaminated with waste water treatment plant effluent, while other faecal indicator bacteria, e.g. from animals or adapted environmental strains, were detected in the sediment of the bay. Another study used the method not only for identification, but as a typing tool. The authors reported the application of MALDI-TOF profiling for discrimination of closely related *E. coli* strains found in environment and tracking their respective sources (Siegrist *et al.*, 2007). Niyompanich and co-workers investigated MALDI-TOF MS whole cell profile spectra for biomarkers, which could differentiate strains from different sources, i.e. environmental and clinically relevant strains of *Burkholderia pseudomallei*, a highly pathogenic bacterium which is the causative agent of melioidosis (Niyompanich *et al.*, 2014). The group applied a biomarker approach to detect particular source-specific mass signals, which were helpful to distinguish environmental- versus clinical-derived strains.

MALDI-TOF MS and Biological Resource Centres (BRC): the example of the Collection de l'Institut Pasteur (CIP)

General considerations on Biological Resource Centres

In March 2012, the Organisation for Economic Co-operation and Development (OECD) published a manual of recommendations entitled 'OECD best practice guidelines for Biological Resource Centres'. It included general instructions for the acquisition, maintenance and provision of biological materials and on the management of Biological Resource Centres (BRC), with the aim of ensuring that biological materials delivered to the international community of scientists and industry are of the highest standard and authentic.

Consequently, the roles of a BRC are:

- to conserve biodiversity by regular enrichment;
- to make available biological materials for R and D purposes in the fields of science, medicine, education, industry and environment;
- to centralize documented resources, to facilitate exchanges;
- to make available information and biological materials through a catalogue and/or website;
- to harmonize and standardize the techniques used;
- to ensure traceability and compliance of the biological materials by the development of a performant quality system
- to create a job opportunity for joint working bases;
- to serve as a reference (controlled and certified database).

The World Federation for Culture Collections (WFCC) (http://www.wfcc.info/) has developed an international database on biological resources worldwide named the World Data Centre for Microorganisms (WDCM). 692 culture collections from 71 countries are currently counted. In France, 38 culture collections are registered in WFCC representing 86,000 holdings (http://www.wfcc.info/ccinfo/home/). At European level, the European Culture Collections' Organisation (ECCO) (https://www.eccosite.org/) was established in 1981. It currently comprises 61 members from 22 European countries. The total holdings of the collections number over 350,000 strains representing yeasts, filamentous fungi, bacteria, archaea, algae and protozoa etc. The aim of the organization is to promote collaboration and exchange of ideas and information about all aspects of culture collection activity. Thus, through the European project EMbaRC (http://www.embarc.eu/) gathering 10 European

BRCs, MALDI-TOF mass spectrometry was used to compare holdings in three of them (CIP from France, CECT from Spain, and DSMZ from Germany) allowing the achievement of an inter-laboratory analysis of 451 equivalent strains of 81 species and 15 subspecies of the genus *Lactobacillus*, 25 species and three subspecies of the family *Leuconostocaceae* and 22 species and two subspecies of the genus *Pseudomonas* (http://www.embarc.eu/deliverables/EMbaRC_D.JRA2.1.3_D15.15_MALDI-TOF.pdf).

Basic activities of a BRC are enrichment with original material, production/preservation, control, and distribution. They are carried out in the context of a proven quality system management, and the use of dedicated software.

The preservation techniques used should retain the full potential of the biological material and ensure its consistency between centres supplying it. This will help to provide a reliable basis for research and development in different laboratories. Bacterial strains are preserved as far as possible through two ways (freeze-drying and deep-freezing at −80°C). If bacteria are recalcitrant to freeze-drying they are preserved using deep-freezing at −80°C and liquid nitrogen.

During the production step, control testing is used to check the conformity of the biological material. A strain is considered as 'conform to specifications' when the results of the controls matched perfectly to those indicated by the depositary laboratory. For the description of a new species, there is an obligation for scientists to deposit the type strain of the new species in two different BRCs in two different countries and obligation for these BRCs to give a certificate of deposit linked with the checks carried out.

The strains delivered by BRCs are ordered by research laboratories, universities, and private and public laboratories. Distribution is ensured in France and throughout the world, according to safety norms for health and environment and in accordance with the regulations and laws in force allowing the traceability of all operations. The Nagoya Protocol (NP), which entered into force on 12 October 2014, especially requires a perfect traceability. Indeed, it is an international agreement which aims at sharing the benefits arising from the utilization of genetic resources in a fair and equitable way.

Strain identification within the Collection de l'Institut Pasteur (CIP)

BRCs are regarded as reference sites (e.g. issuance of a certificate of deposit for strains representing new species), which requires significant expertise in identification. In the past, identification of deposited bacterial strains the work necessary for secure species identification was particularly time-consuming and difficult because identification was based on biochemical characteristics. Nowadays, the molecular and mass spectrometry methods significantly simplify and improve bacterial identification.

In the CIP, more than 15,000 bacterial strains are preserved, representing 4840 different species; each year around 150 new deposits are received at the CIP and freeze-dried or frozen for preservation. In addition, the batch renewal of almost as many strains already present in the CIP is necessary, as a result of their distribution. Control of the strain's species name is an essential step before the strains become available for the distribution. The currently used standard methods for strain identification within the CIP are 16S rDNA sequencing and MALDI-TOF mass spectrometry. The choice of these two techniques was done because 16S rDNA sequencing is an established molecular criterion for species delineation

(Stackebrandt *et al.*, 2002) and MALDI-TOF mass spectrometry is recognized as a fast and reliable microorganism identification method (Mellmann *et al.*, 2009; Panda *et al.*, 2015). Other techniques like multi locus sequence analysis and whole genome sequencing are used when necessary, i.e. when the systematic resolving power of both standard methods is not high enough.

It is interesting to notice that in recent years, MALDI-TOF MS was used as one of the standard techniques for description of novel species. Thus, the description of the two novel species *Streptococcus tangierensis* sp. nov. and *Streptococcus cameli* sp. nov. relies on both the results obtained with the MALDI-TOF MS and biochemical tests (Kadri *et al.*, 2015). In 2013, Benejat and colleagues demonstrated that MALDI-TOF MS can be used as an alternative method to 16S rDNA sequencing for phylogeny and can lead to the discovery of new *Campylobacter* species (Benejat *et al.*, 2013). Regarding the study performed by Gomila *et al.* (2014), MALDI-TOF MS analyses of the *Achromobacter* species provided an alternative method that could be correlated with the resolution of the *Achromobacter* species by selected housekeeping gene sequence analyses. Thereby, in the upcoming years adding MALDI-TOF MS analyses will become unavoidable in studies and publications describing novel species.

The bacterial strains newly incorporated into the CIP are subjected to 16S rRNA gene sequencing and MALDI-TOF mass spectrometry. The formerly incorporated bacterial strains, which were only identified on the basis of their biochemical characteristics, are gradually also checked the same way during batch renewal to establish a comprehensive data fundament. In our routine work, these two techniques, with hindsight of several years, have been demonstrated to have a very good concordance for bacterial identification. In some cases, MALDI-TOF MS profiles are showing a higher resolution power. Thus, for instance *Citrobacter freundii*, *Citrobacter braakii*, and *Citrobacter youngae* which are showing more than 98% sequence similarity among their 16S rRNA gene sequences and are phylogenetically closely related as seen in Fig. 7.1, are well separated in the tree based on MALDI-TOF mass spectra (Fig. 7.2). In few other cases, 16S rDNA sequencing is superior, e.g. for differentiation of *Bacteroides dorei* from *Bacteroides vulgatus* and *Bacteroides ovatus* from *Bacteroides xylanisolvens* (Pedersen *et al.*, 2013).

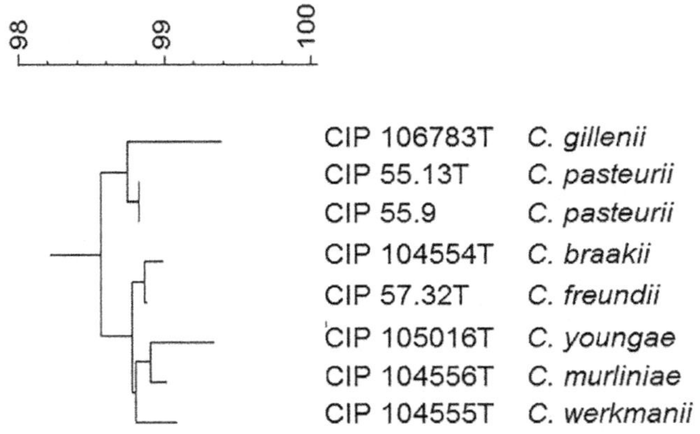

Figure 7.1 Phylogenetic tree based on 16S rRNA gene sequences showing the relationships between some *Citrobacter* species. Relationship is given in% DNA sequence identity.

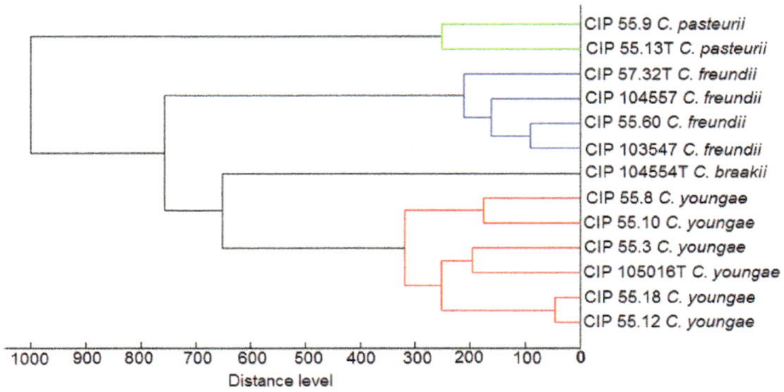

Figure 7.2 MSP based dendrogram of *Citrobacter* strains. The similarity dendrogram has been calculated from reference spectra (MSPs) calculated by the MALDI Biotyper software and using the according dendrogram functionality of the software with standard settings.

The CIP has been established as early as in 1892 by Doctor Binot and a number of strains deposited in the early twentieth century have only been identified using phenotypical and biochemical methods. These isolates need to be re-identified before distribution. In this case, MALDI TOF MS is used as first-line method for the fast re-identification at genus or species level.

The routine work in the CIP demonstrates the power of MALDI-TOF MS for bacterial identification for systematic purposes. The technique meets both the criteria for quality of identification, speed and low cost of consumables. Moreover, it is important that reference BRCs are involved in the updating of the MALDI-TOF MS database because the type strains of new species are deposited in such collections before publication.

References

Ali, Z., Cousin, S., Frühling, A., Brambilla, E., Schumann, P., Yang, Y., and Stackebrandt, E. (2009). *Flavobacterium rivuli* sp. nov., *Flavobacterium subsaxonicum* sp. nov., *Flavobacterium swingsii* sp. nov. and *Flavobacterium reichenbachii* sp. nov., isolated from a hard water rivulet. Int. J. Syst. Evol. Microbiol. *59*, 2610–2617.

Behrendt, U., Schumann, P., Meyer, J.-M., and Ulrich, A. (2009). *Pseudomonas cedrina* subsp. *fulgida subsp. nov.*, a fluorescent bacterium isolated from the phyllosphere of grasses; emended description of *Pseudomonas cedrina* and description of *Pseudomonas cedrina* subsp. *cedrina subsp. nov.* Int. J. Syst. Evol. Microbiol. *59*, 1331–1335.

Behrendt, U., Schumann, P., Stieglmeier, M., Pukall, R., Augustin, J., Spröer, C., Schwendner, P., Moissl-Eichinger, C., and Ulrich, A. (2010). Characterization of heterotrophic nitrifying bacteria with respiratory ammonification and denitrification activity--description of *Paenibacillus uliginis* sp. nov., an inhabitant of fen peat soil and *Paenibacillus purispatii* sp. nov., isolated from a spacecraft assembly clean room. Syst. Appl. Microbiol. 33, 328–336.

Benejat, L., Gravet, A., Sifré, E., Ben Amor, S., Quintard, B., Mégraud, F., and Lehours, P. (2014). Characterization of a *Campylobacter* fetus-like strain isolated from the faeces of a sick leopard tortoise (*Stigmochelys pardalis*) using matrix-assisted laser desorption/ionization time of flight as an alternative to bacterial 16S rDNA phylogeny. Lett. Appl. Microbiol. *58*, 338–343.

Bittar, F., Keita, M.B., Lagier, J.-C., Peeters, M., Delaporte, E., and Raoult, D. (2014). *Gorilla gorilla gorilla* gut: a potential reservoir of pathogenic bacteria as revealed using culturomics and molecular tools. Sci. Rep. *4*, 7174.

Cassir, N., Papazian, L., Fournier, P.-E., Raoult, D., and La Scola, B. (2015). Insights into bacterial colonization of intensive care patients' skin: the effect of chlorhexidine daily bathing. Eur. J. Clin. Microbiol. Infect. Dis. *34*, 999–1004.

Dieckmann, R., Graeber, I., Kaesler, I., Szewzyk, U., and von Döhren, H. (2005). Rapid screening and dereplication of bacterial isolates from marine sponges of the sula ridge by intact-cell-MALDI-TOF mass spectrometry (ICM-MS). Appl. Microbiol. Biotechnol. *67*, 539–548.

Dieckmann, R., Strauch, E., and Alter, T. (2010). Rapid identification and characterization of *Vibrio* species using whole-cell MALDI-TOF mass spectrometry. J. Appl. Microbiol. *109*, 199–211.

Donohue, M.J., Best, J.M., Smallwood, A.W., Kostich, M., Rodgers, M., and Shoemaker, J.A. (2007). Differentiation of *Aeromonas* isolated from drinking water distribution systems using matrix-assisted laser desorption/ionization-mass spectrometry. Anal. Chem. 79, 1939–1946.

Dubourg, G., Lagier, J.C., Armougom, F., Robert, C., Hamad, I., Brouqui, P., and Raoult, D. (2013). The gut microbiota of a patient with resistant tuberculosis is more comprehensively studied by culturomics than by metagenomics. Eur. J. Clin. Microbiol. Infect. Dis. *32*, 637–645.

Dubourg, G., Lagier, J.C., Robert, C., Armougom, F., Hugon, P., Metidji, S., Dione, N., Dangui, N.P.M., Pfleiderer, A., Abrahao, J., *et al.* (2014). Culturomics and pyrosequencing evidence of the reduction in gut microbiota diversity in patients with broad-spectrum antibiotics. Int. J. Antimicrob. Agents *44*, 117–124.

Dybwad, M., Granum, P.E., Bruheim, P., and Blatny, J.M. (2012). Characterization of airborne bacteria at an underground subway station. Appl. Environ. Microbiol. 78, 1917–1929.

Eddabra, R., Prévost, G., and Scheftel, J.-M. (2012). Rapid discrimination of environmental *Vibrio* by matrix-assisted laser desorption ionization time-of-flight mass spectrometry. Microbiol. Res. *167*, 226–230.

Emami, K., Askari, V., Ullrich, M., Mohinudeen, K., Anil, A.C., Khandeparker, L., Burgess, J.G., and Mesbahi, E. (2012). Characterization of bacteria in ballast water using MALDI-TOF mass spectrometry. PLoS One 7, e38515.

Emami, K., Hack, E., Nelson, A., Brain, C.M., Lyne, F.M., Mesbahi, E., Day, J.G., and Caldwell, G.S. (2015). Proteomic-based biotyping reveals hidden diversity within a microalgae culture collection: An example using *Dunaliella*. Sci. Rep. *5*, 10036.

Erler, R., Wichels, A., Heinemeyer, E.-A., Hauk, G., Hippelein, M., Reyes, N.T., and Gerdts, G. (2015). VibrioBase: A MALDI-TOF MS database for fast identification of *Vibrio* spp. that are potentially pathogenic in humans. Syst. Appl. Microbiol. *38*, 16–25.

Fox, K., Fox, A., Elssner, T., Feigley, C., and Salzberg, D. (2010). MALDI-TOF mass spectrometry speciation of staphylococci and their discrimination from micrococci isolated from indoor air of schoolrooms. J. Environ. Monit. *12*, 917–923.

Gehring, I., Wensing, A., Gernold, M., Wiedemann, W., Coplin, D.L., and Geider, K. (2014). Molecular differentiation of *Pantoea stewartii* subsp. *indologenes* from subspecies *stewartii* and identification of new isolates from maize seeds. J. Appl. Microbiol. *116*, 1553–1562.

Ghyselinck, J., Van Hoorde, K., Hoste, B., Heylen, K., and De Vos, P. (2011). Evaluation of MALDI-TOF MS as a tool for high-throughput dereplication. J. Microbiol. Methods *86*, 327–336.

Gomila, M., Prince-Manzano, C., Svensson-Stadler, L., Busquets, A., Erhard, M., Martínez, D.L., Lalucat, J., and Moore, E.R.B. (2014). Genotypic and phenotypic applications for the differentiation and species-level identification of achromobacter for clinical diagnoses. PLoS One *9*, e114356.

Gouba, N., Raoult, D., and Drancourt, M. (2014). Eukaryote culturomics of the gut reveals new species. PLoS One 9, e106994.

Hausdorf, L., Mundt, K., Winzer, M., Cordes, C., Fröhling, A., Schlüter, O., and Klocke, M. (2013). Characterization of the cultivable microbial community in a spinach-processing plant using MALDI-TOF MS. Food Microbiol. 34, 406–411.

Hazen, T.H., Martinez, R.J., Chen, Y., Lafon, P.C., Garrett, N.M., Parsons, M.B., Bopp, C.A., Sullards, M.C., and Sobecky, P.A. (2009). Rapid identification of *vibrio parahaemolyticus* by whole-cell matrix-assisted laser desorption ionization-time of flight mass spectrometry. Appl. Environ. Microbiol. 75, 6745–6756.

Hundenborn, J., Thurig, S., Kommerell, M., Haag, H., and Nolte, O. (2013). Severe wound infection with *photobacterium damselae* ssp. *damselae* and *vibrio harveyi*, following a laceration injury in marine environment: a case report and review of the literature. Case Rep. Med. *2013*, 610632.

Kadri, Z., Vandamme, P., Ouadghiri, M., Cnockaert, M., Aerts, M., Elfahime, E.M., Farricha, O.E., Swings, J., and Amar, M. (2015). *Streptococcus tangierensis* sp. nov. and *Streptococcus cameli* sp. nov., two novel *Streptococcus* species isolated from raw camel milk in Morocco. Antonie Van Leeuwenhoek *107*, 503–510.

Lagier, J.-C., Armougom, F., Million, M., Hugon, P., Pagnier, I., Robert, C., Bittar, F., Fournous, G., Gimenez, G., Maraninchi, M., *et al.* (2012a). Microbial culturomics: paradigm shift in the human gut microbiome study. Clin. Microbiol. Infect. *18*, 1185–1193.

Lagier, J.-C., Million, M., Hugon, P., Armougom, F., and Raoult, D. (2012b). Human gut microbiota: repertoire and variations. Front. Cell. Infect. Microbiol. *2*, 136.

Levican, A., Rubio-Arcos, S., Martinez-Murcia, A., Collado, L., and Figueras, M.J. (2015). *Arcobacter ebronensis* sp. nov. and *Arcobacter aquimarinus* sp. nov., two new species isolated from marine environment. Syst. Appl. Microbiol. *38*, 30–35.

Madsen, A.M., Zervas, A., Tendal, K., and Nielsen, J.L. (2015). Microbial diversity in bioaerosol samples causing ODTS compared to reference bioaerosol samples as measured using Illumina sequencing and MALDI-TOF. Environ. Res. *140*, 255–267.

Malainine, S.M., Moussaoui, W., Prévost, G., Scheftel, J.-M., and Mimouni, R. (2013). Rapid identification of *Vibrio parahaemolyticus* isolated from shellfish, sea water and sediments of the Khnifiss lagoon, Morocco, by MALDI-TOF mass spectrometry. Lett. Appl. Microbiol. *56*, 379–386.

Martínez-Murcia, A.J., Saavedra, M.J., Mota, V.R., Maier, T., Stackebrandt, E., and Cousin, S. (2008). *Aeromonas aquariorum* sp. nov., isolated from aquaria of ornamental fish. Int. J. Syst. Evol. Microbiol. *58*, 1169–1175.

Mellmann, A., Bimet, F., Bizet, C., Borovskaya, A.D., Drake, R.R., Eigner, U., Fahr, A.M., He, Y., Ilina, E.N., Kostrzewa, M., *et al.* (2009). High interlaboratory reproducibility of matrix-assisted laser desorption ionization-time of flight mass spectrometry-based species identification of nonfermenting bacteria. J. Clin. Microbiol. *47*, 3732–3734.

Munoz, R., López-López, A., Urdiain, M., Moore, E.R.B., and Rosselló-Móra, R. (2011). Evaluation of matrix-assisted laser desorption ionization-time of flight whole cell profiles for assessing the cultivable diversity of aerobic and moderately halophilic prokaryotes thriving in solar saltern sediments. Syst. Appl. Microbiol. *34*, 69–75.

Ng, H.J., López-Pérez, M., Webb, H.K., Gomez, D., Sawabe, T., Ryan, J., Vyssotski, M., Bizet, C., Malherbe, F., Mikhailov, V.V., *et al.* (2014). *Marinobacter* salarius sp. nov. and *Marinobacter similis* sp. nov., isolated from sea water. PLoS One *9*, e106514.

Niyompanich, S., Jaresitthikunchai, J., Srisanga, K., Roytrakul, S., and Tungpradabkul, S. (2014). Source-identifying biomarker ions between environmental and clinical *Burkholderia pseudomallei* using whole-cell matrix-assisted laser desorption/ionization time-of-flight mass spectrometry (MALDI-TOF MS). PLoS One *9*, e99160.

Oberbeckmann, S., Wichels, A., Maier, T., Kostrzewa, M., Raffelberg, S., and Gerdts, G. (2011). A polyphasic approach for the differentiation of environmental *Vibrio* isolates from temperate waters. FEMS Microbiol. Ecol. *75*, 145–162.

Panda, A., Ghosh, A.K., Mirdha, B.R., Xess, I., Paul, S., Samantaray, J.C., Srinivasan, A., Khalil, S., Rastogi, N., and Dabas, Y. (2015). MALDI-TOF mass spectrometry for rapid identification of clinical fungal isolates based on ribosomal protein biomarkers. J. Microbiol. Methods *109*, 93–105.

Pedersen, R.M., Marmolin, E.S., and Justesen, U.S. (2013). Species differentiation of *Bacteroides dorei* from *Bacteroides vulgatus* and *Bacteroides ovatus* from *Bacteroides xylanisolvens* – back to basics. Anaerobe *24*, 1–3.

Pfleiderer, A., Lagier, J.-C., Armougom, F., Robert, C., Vialettes, B., and Raoult, D. (2013). Culturomics identified 11 new bacterial species from a single anorexia nervosa stool sample. Eur. J. Clin. Microbiol. Infect. Dis. *32*, 1471–1481.

Sauer, S., Freiwald, A., Maier, T., Kube, M., Reinhardt, R., Kostrzewa, M., and Geider, K. (2008). Classification and identification of bacteria by mass spectrometry and computational analysis. PLoS One *3*, e2843.

Siegrist, T.J., Anderson, P.D., Huen, W.H., Kleinheinz, G.T., McDermott, C.M., and Sandrin, T.R. (2007). Discrimination and characterization of environmental strains of *Escherichia coli* by matrix-assisted laser desorption/ionization time-of-flight mass spectrometry (MALDI-TOF-MS). J. Microbiol. Methods *68*, 554–562.

Stackebrandt, E., Frederiksen, W., Garrity, G.M., Grimont, P.A.D., Kämpfer, P., Maiden, M.C.J., Nesme, X., Rosselló-Mora, R., Swings, J., Trüper, H.G., *et al.* (2002). Report of the ad hoc committee for the re-evaluation of the species definition in bacteriology. Int. J. Syst. Evol. Microbiol. *52*, 1043–1047.

Thevenon, F., Regier, N., Benagli, C., Tonolla, M., Adatte, T., Wildi, W., and Poté, J. (2012). Characterization of fecal indicator bacteria in sediments cores from the largest freshwater lake of Western Europe (Lake Geneva, Switzerland). Ecotoxicol. Environ. Saf. *78*, 50–56.

Uhlik, O., Strejcek, M., Junkova, P., Sanda, M., Hroudova, M., Vlcek, C., Mackova, M., and Macek, T. (2011). Matrix-assisted laser desorption ionization (MALDI)-time of flight mass spectrometry- and MALDI biotyper-based identification of cultured biphenyl-metabolizing bacteria from contaminated horseradish rhizosphere soil. Appl. Environ. Microbiol. *77*, 6858–6866.

Wensing, A., Zimmermann, S., and Geider, K. (2010). Identification of the corn pathogen *Pantoea stewartii* by mass spectrometry of whole-cell extracts and its detection with novel PCR primers. Appl. Environ. Microbiol. *76*, 6248–6256.

Wensing, A., Gernold, M., and Geider, K. (2012). Detection of *Erwinia* species from the apple and pear flora by mass spectroscopy of whole cells and with novel PCR primers. J. Appl. Microbiol. *112*, 147–158.

Zhang, L., Vranckx, K., Janssens, K., and Sandrin, T.R. (2015). Use of MALDI-TOF mass spectrometry and a custom database to characterize bacteria indigenous to a unique cave environment (Kartchner Caverns, AZ, USA). J. Vis. Exp. e52064.

The World of Nucleic Acid-based Mass Spectrometry for Microbial and Viral Detection

8

Christiane Honisch

Abstract

Matrix-assisted laser desorption ionization time-of-flight mass spectrometry (MALDI-TOF MS) has been applied to nucleic acid based research and specifically sequencing for over 25 years with increasing clinical use today. In contrast to whole cell MS, where cultured material is the source for the identification of a microbial sample based on a proteomic fingerprint, nucleic acid MS based microbial detection, identification and characterization uses PCR amplifications of specific genomic target regions of interest followed by the detection of single nucleotide polymorphisms (SNPs) via primer extension reactions or followed by comparative sequence analysis of mass peak pattern generated via *in vitro* transcription and base-specific cleavage. Applications include the detection and identification of both microbial and viral pathogens, tracking of transmissions and the characterization of genetic heterogeneity by variant analysis. Multiplexing levels of up to 40–60 targets per reaction surpass real-time PCR applications, but fall short when compared to the potential of next-generation sequencing. The unique feature of mass spectrometry (MS) though is the detection of various biomarker molecules including proteins, lipids, small molecules, carbohydrates and nucleic acids on one single platform.

Introduction

Mass spectrometry (MS) is the superior technology when it comes to the detection of various biomarker molecules including proteins, lipids, small molecules, carbohydrates and nucleic acids on a single platform. Beyond experimental research the technology has thus entered public health and clinical laboratories. Especially, matrix-assisted laser desorption ionization time-of-flight mass spectrometry (MALDI-TOF MS) of whole cells has been broadly accepted by clinical microbiology laboratories throughout the world as an innovative tool for bacterial species identification (van Belkum *et al.*, 2012). Within the last 2 years whole cell analysis of cultured organisms on the bioMeriéux Vitek MS and on the Bruker MALDI Biotyper as well as nucleic acid based genotyping assays for Factor V Leiden and Factor II on the Agena IMPACT Dx mass spectrometer (former Sequenom MassARRAY) received FDA approval. PCR product analysis on the Abbott Iridica MS (earlier platform Ibis Biosciences ESI MS T5000) got CE-marked in late 2014.

All of the assay-specific commercial platforms use a soft ionization or non-fragmenting electrospray technology (ESI) (Fenn *et al.*, 1989) or a matrix carrier (MALDI) in combination with a laser ion source (Karas and Hillenkamp, 1988) to detect the molecular mass to charge ratio (m/z) of the intact analyte – its intrinsic physical property. This means no fluorescent or radio labelling for nucleic acid specific detection assays. The accuracies and resolution of the benchtop mass spectrometers mentioned above allow for the detection of hundreds of nucleic acid specific signals per spectrum, which translates into very competitive multiplexing levels when compared to fluorescent four or five dye based systems. Medium to high-throughput data acquisition of up to 384 sample spectra per hour can be obtained.

These MS platforms are open systems and support biochemistries for nucleic acid based detection with a discrimination at a single nucleotide level. Biochemistries are based on PCR amplifications of the target region of interest followed by the direct detection of the PCR products (see Fig. 8.4a), the identification of single nucleotide polymorphisms via primer extension (see Fig. 8.1) or comparative sequence analysis of mass peak pattern generated via *in vitro* transcription and base-specific cleavage (see Fig. 8.3). Additional variations of these biochemistries like PCR in combination with restriction digests have been described (Hong *et al.*, 2008) (see Fig. 8.4b). Genomic fields of application are target identification, genotyping/mutation and variant detection, comparative sequencing, methylation analysis and quantitative expression profiling.

The preparation of the analyte is highly amenable to automation all the way from sample preparation through PCR amplification and clean-up. A fully automated platform has not been commercialized but procedures on existing fluorescent based platforms could be used as feasible models for development and implementation.

The clinical microbiology lab continues to rely heavily on traditional microbiological methods for the taxonomic classification and identification of bacteria including culture, phenotyping and biochemical methods to identify infectious agents present in clinical specimens. These procedures are validated but laborious and can be subjective. A breakthrough has been the successful implementation of whole cell MALDI-TOF MS as a routine identification tool of pure cultures about a decade ago. On the other hand, the extremely low limit of detection (1 to 10 copies of target) and rapid results of molecular methods have let to proposed changes in the definition of culture as a gold standard for the detection and identification of bacteria and fungi in clinical specimens, especially for those that are difficult to culture. Nucleic acid based procedures can generate a robust readout from the bacterial sample, as the genome remains largely stable over time and is resistant to environmental changes during sample transport (Sauer and Kliem 2010). Sequence analysis of microbial genomes has the potential to quickly and accurately identify antibiotic resistance that is mediated by gene mutations, mobile genetic elements or virulence factors. It is important to recognize though that these approaches only confirm the presence of a nucleic acid target and do not prove the presence of a viable organism as detected by culture. For patient diagnostics it is important that molecular assays are interpreted in the context of clinical presentation. They should not be used as a test of cure as the presence of nucleic acid does not always correlate with clinical illness (Buchan and Ledeboer, 2014).

Clinical virology on the other hand has shifted to molecular based tests due to increased sensitivity, specificity and a faster turn-around time when compared to viral cultures.

In general, molecular diagnostics of infectious agents is based on the specific recovery and detection of certain genus and/or species specific genomic fragments inclusive for the infectious agent of interest and exclusive for additional species in the sample flora and nearest phylogenetic neighbours. Further identification and characterization is based on the sequence analysis of additional genomic regions like 16S rDNA loci and today even of whole genomes. DNA microheterogeneities form the basis of most epidemiological typing tools like multilocus sequence typing (MLST). Low levels of an infectious agent can be detected quantitatively in complex samples due to the efficient and specific amplification of a target region by PCR. DNA or RNA purification from the clinical specimen is usually included before sample preparation.

Nucleic acid based MS has successfully been used for the identification of both microbial and viral pathogens, tracking of transmissions and the characterization of genetic heterogeneity by variant analysis and genomic comparative sequencing (Ganova-Raeva and Khudyakov, 2013).

Nucleic Acid MALDI-TOF MS

Standard molecular methods for the characterization of bacteria like genotyping, spoligotyping, multi-locus sequence typing (MLST), 16S rDNA based typing and restriction fragment length polymorphism analysis as well as genotyping and comparative sequencing of RNA and DNA viruses and bacteria have been implemented on MALDI-TOF MS.

All applications use specific or broad range PCR primers in combination with different post-PCR biochemistries to substitute or improve the classical molecular approaches.

MALDI-TOF MS has been applied to sequencing for over 25 years. Limitations like salt adduct formation and analyte fragmentation have been overcome at least for small nucleic acid fragments. Robust biochemistries have been developed. Owing to their negatively charged phosphate backbone, nucleic acids are susceptible to adduct formation with cations – predominantly sodium and potassium from the surrounding reaction mixtures. Adduct formation lowers sensitivity and analytical accuracy. Streptavidin–biotin solid phase clean-up (Tang *et al.*, 1995) and ion exchange procedures (Nordhoff *et al.*, 1992) preferably for ammonium have been implemented. Ammonium is a volatile cation in the gas phase of the mass spectrometer released as ammonia leaving the nucleic acid analyte molecules as free acids for detection.

Based on the chemical nature of nucleic acids, protonation of the nucleobases A or G during the MALDI process induces polarization of the N-glycosidic bond between the sugar and nucleobase, which can finally result in nucleobase elimination. Subsequent to depurination, fragmentation can occur via backbone cleavage. It has been demonstrated that ribonucleic acid (RNA) is less susceptible to fragmentation under MALDI conditions than deoxyribonucleic acid (DNA) due to the lack of the 2′ hydroxy group in the ribose sugar moiety (Nordhoff *et al.*, 1993; Kirpekar *et al.*, 1994).

A milestone was the introduction of a nucleic acid favourable matrix (3-hydropicolinic acid (3-HPA)) (Wu *et al.*, 1993), some of its derivatives and an optimal matrix-to-analyte ratio.

Solid-phase Sanger DNA sequencing and the detection of the extended products by MALDI-TOF MS was shown by Köster *et al.* in 1996. The approach is limited by the exponential decay in sensitivity of MALDI-TOF MS and the resolution of a conventional axial-TOF MS instrument, which translates into a maximum differentiable product length of about 100 bp. *De novo* Sanger DNA sequencing with read length of 800–1000 bp on electrophoresis or capillary platforms is thus impracticable by MALDI-TOF MS. However, the short sequencing approach is particularly suitable for the high throughput detection of one or multiple mutations (Jurinke *et al.*, 2004) or for the resolution of fragments that cannot be reliably analysed in Sanger sequencing in the standard electrophoretic manner due to compression or false termination.

The method evolved into an assay design and data analysis software assisted product. The initially streptavidin coated bead-based primer extension reaction improved to a homogeneous primer extension assay with specific deoxyribonucleotide (dNTP)/dideoxynucleotide (ddNTP) stop mixes for multiplexed SNP analysis (Fig. 8.1a) (Storm *et al.*, 2002). Recent developments of the biochemistry into a PCR/single base primer extension assay (MassARRAY iPLEX biochemistry (Agena Bioscience)), which utilizes mass modified dideoxynucleotide terminators, allow for the simultaneous detection and differentiation of extension products of all four nucleotides (Fig. 8.1b). Higher multiplexing of 40–60 heterozygous loci and single nucleotide polymorphisms as well as insertions and deletions in one reaction can be achieved.

All reactions are performed in the same reaction plate by liquid handling with the added quality advantage of traceability of samples and reduced human error throughout the entire process. Products are purified through the addition of ion-exchange resin and subsequently dispensed on a chip array that is preloaded with a modified 3-HPA matrix preparation.

The latest expansion of the PCR/extension based biochemistry allows for ultrasensitive detection of mutations down to 0.125% of a mutant in the background of wildtype as demonstrated for the detection of somatic cancer mutations. The extension reaction of this approach utilizes a single mutation specific chain terminator labelled with a moiety for solid phase capture (Fig. 8.1c). Captured, washed, and eluted products are interrogated for mass

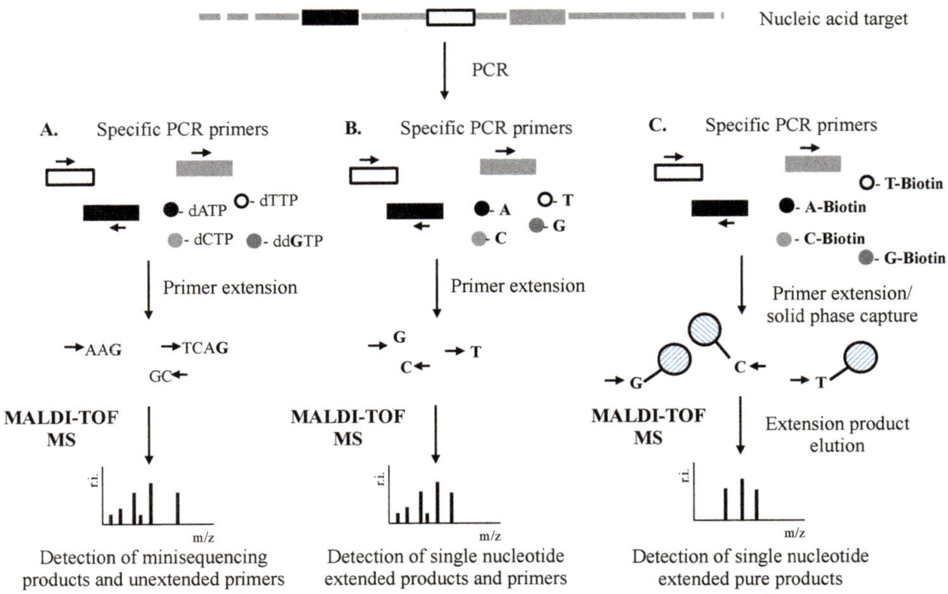

Figure 8.1 Schematic workflow of PCR/Primer Extension MALDI-TOF MS assays. (A) PCR and primer extension with dNTP/ddNTP stop mixes; (B) PCR and primer extension with mass modified deoxyribonucleotides; (C) PCR and primer extension with biotinylated dexoyribonucleotides, bead capture and purification.

and mutational genotypes using MALDI-TOF MS. This new method, when compared to next generation sequencing platforms, offers a relatively simple workflow with quick turnaround time and minimal sample input requirements (Nygren *et al.*, 2013).

The multiplexing capacity of the MALDI-TOF MS PCR/primer extension assay has been used for genotyping and the detection of RNA and DNA viruses with a throughput that supports large-scale epidemiological research studies.

The technology enabled for example the detection of hepatitis C (HCV) genotype specific point mutations with clinical importance for disease outcome and response to therapy (Ilina *et al.*, 2005), the differentiation of high and low-risk human papillomavirus (HPV) genotypes in different tissue types as a potential tag for tracking certain types of cancer (Stenmark *et al.*, 2013; Yang *et al.*, 2004) and the characterization of SARS coronavirus (SARS-CoV) sequence variations (Liu *et al.*, 2005). Furthermore, the approach enables the detection of several viruses in a single sample as shown for the differentiation of human herpesviruses in a wide variety of archival biological specimens (Sjöholm *et al.*, 2008) and for the detection of eight human enteric viruses (hepatitis E, coxsackievirus, poliovirus, astrovirus, norovirus, echovirus and reovirus) (Piao *et al.*, 2012).

The concordance rate between MALDI-TOF MS and reference methods like oligonucleotide microarray, real-time PCR and dideoxy sequencing is generally high and the detection limits are comparable. For viruses, e.g. human enteric viruses, sensitivities from 100 to 1000 copies per reaction have been described. A distinct advantage of the multiplexing capabilities is that it is possible to extend existing assays by adding new type-specific primers (Cobo, 2013). When compared to a technology like dideoxy sequencing, data acquisition times are much faster requiring a few seconds per sample on a MALDI-TOF MS instrument. Twenty-four, 96 or 384 samples can be analysed in one run.

When SNP panels provide sufficient discriminatory power and resolution, PCR/single primer extension MS can be applied to the characterization of bacteria and the study of their relatedness, diversity and spread. Examples are a 16 SNP-based MRSA characterization assay (Symis *et al.*, 2011), a MLST-style 14 SNP based genotyping assay for *Neisseria gonorrhoeae* (Trembizki *et al.*, 2013) and a MLST-style 20 SNP based assay for *Pseudomonas aeruginosa* strain genotyping in cystic fibrosis patients (Syrmis *et al.*, 2014).

In mixtures, frequencies of extension products of all four nucleotides can be determined by calculating the areas of the peaks associated with the corresponding products. An application of this quantitative approach on viruses was demonstrated in a study supporting vaccine control. Viral quasispecies of the mumps virus were determined between Jeryl Lynn substrains in live, attenuated mumps/measles vaccine based on five distinct nucleotide positions in the viral genome. Feasibility was shown in reference with the existing QC methodology used by the Federal Drug Administration (FDA) (Amexis *et al.*, 2001).

Many mycobacterial species, including *Mycobacterium tuberculosis*, grow extremely slowly in the laboratory and require 3–8 weeks of incubation on solid medium or at least 2 weeks in a radiometric liquid culture system (BACTEC). This slow growth often leads to a delay in tuberculosis (TB) diagnosis (Soini and Musser, 2001). Molecular amplification methods have a distinct advantage in TB diagnosis as they allow for the characterization of *M. tuberculosis*, its Complex and linages as well as antibiotic resistance before culture results are available. Different rapid nucleic acid MALDI-TOF MS applications provide a suite of assays for the characterization of this pathogen.

M. tuberculosis is a member of the *M. tuberculosis* Complex (MTBC) (Tsukamura *et al.*,

1985), which comprises eight closely related bacterial species with distinct host tropism including the human pathogens *M. tuberculosis, M. africanum* and *M. canettii* and the animal-adapted pathogens *M. bovis, M. microti, M. caprae* and *M.pinnipedii* as well as the recently identified species *M. mungi* (Alexander *et al.*, 2010). A 16 SNP assay based on single base extension on MALDI-TOF MS enables simultaneous differentiation of these MTBC species and the characterization of the main phylogenetic lineages (Bouakaze *et al.*, 2011).

A MALDI-TOF based minisequencing method has been developed and applied for the analysis of rifampin (RIF) and isoniazid (INH) multidrug-resistant (MDR) *M. tuberculosis* strains (Ikryannikova *et al.*, 2007). The method uses PCR amplification of the target regions of interest and a subsequent primer extension reaction with dideoxynucleotides for termination of the growing DNA strand from the extension primer with its 3′-end next to or only a few nucleotides before the mutated site (Jurinke *et al.*, 2004). RIF resistance of *M. tuberculosis* is mainly due to point mutations in an 81-bp target region of the *rpo*B gene, which encodes the β–subunit of DNA-dependent RNA polymerase (Ramaswamy and Musser, 1988). INH resistance is associated with mutations in the *kat*G gene encoding catalase-peroxidase KatG, which activates *inh* during its penetration in the cell. Mutations in the promoter region of the *fab*G1 gene encoding 3-ketoacyl ACP reductase form an operon with a downstream *inh*A gene and cause overexpression of inhA, which has also been shown to be involved in INH resistance (Ramaswamy *et al.*, 2003). A study on 100 *M. tuberculosis strains* collected in Moscow in the years 1997–2005 resulted in the nucleic acid based MALDI-TOF MS detection of 91% RIF-resistant and 94% INH-resistant strains when compared to culture as well as an analytical sensitivity of 50 microbial cells per sputum sample and the detection of one mutant cell among 10 wild type cells in a model dilution experiment (Ikryannikova *et al.*, 2007).

Furthermore, a study has shown that spoligotyping of *M. tuberculosis* by analysis of a 43 spacer region, found in a direct repeat region of the genome, can be done by single nucleotide primer extension on MALDI-TOF MS (Honisch *et al.*, 2010). The analysis of these markers by MALDI-TOF MS, which is currently done by a reverse line blot hybridization assay (Kamerbeek *et al.*, 1997), is much more automated and reproducible. Spoligotyping is a very useful tool for the molecular characterization of *M. tuberculosis* strains for epidemiological purposes.

16S rDNA based dideoxy sequencing has been a gold standard for the molecular identification of microbes for over 20 years. 16S rDNA based comparative sequence analysis by MALDI-TOF MS as described below has been applied to the identification of *Mycobacterium* spp. and their differentiation from the infectious agent *M. tuberculosis* (Lefmann *et al.*, 2004; von Wintzingerode *et al.*, 2002). In addition, comparative sequence analysis by MALDI-TOF MS provides an alternative to the minisequencing as described above for the identification of rifampin (RIF) and isoniazid (INH) resistance in *M. tuberculosis* with the addition of discovery of new resistance mutations. An example of single and multiplexed MDR detection by comparative sequencing is shown in Fig. 8.2.

Comparative sequencing of up to 1000 bp PCR products by MALDI-TOF MS has been accomplished by a nucleic acid base-specific-cleavage approach in analogy to the tryptic digestion of proteins with subsequent peptide mapping in protein MS (Stanssens *et al.*, 2004). The principle resembles the original Maxam and Gilbert approach for DNA sequencing (Maxam and Gilbert, 1977), which represents an identification or resequencing

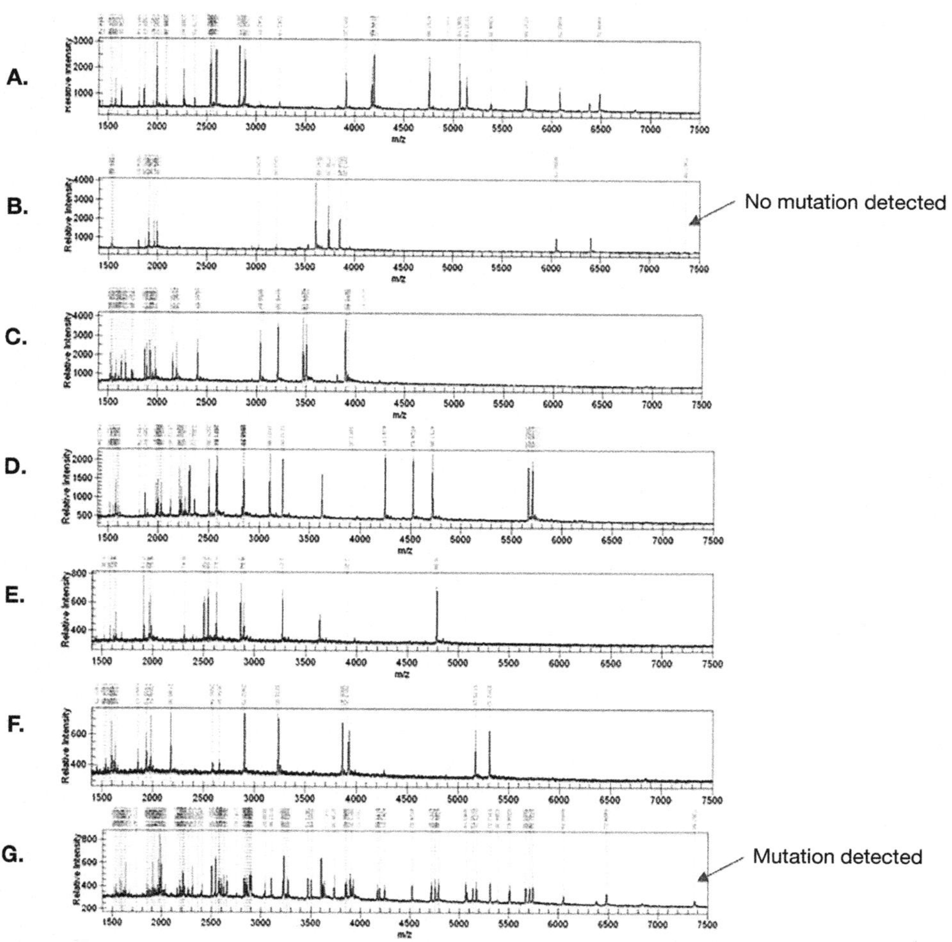

Figure 8.2 MS comparative sequencing of *M. tuberculosis resistance* regions. (A–F) Uniplex of a wildtype *M. tuberculosis* sample; (G) Multiplex of all regions of a mutant *M. tuberculosis* sample.

method that cross-compares an experimentally determined sequence to a known reference sequence.

In comparative sequencing by MALDI-TOF MS (Fig. 8.3), PCR amplicons are generated with gene-specific primers, which are tagged with T7- and SP6-promoter specific sequences. The amplicon is subsequently subjected to transcription by either T7 or SP6 polymerase to obtain transcripts of both strands. Cleavage of the transcripts generates base-specific fragments. Sensitivity of two levels of amplification are achieved, exponential amplification during PCR combined with linear amplification in *in vitro* transcription. During the *in vitro* transcription step dTTP or dCTP are built into the RNA transcript in separate reactions to block subsequent RNase A cleavage at these nucleotides and to achieve nucleotide specific cleavage at riboC and riboU. The four cleavage reactions are

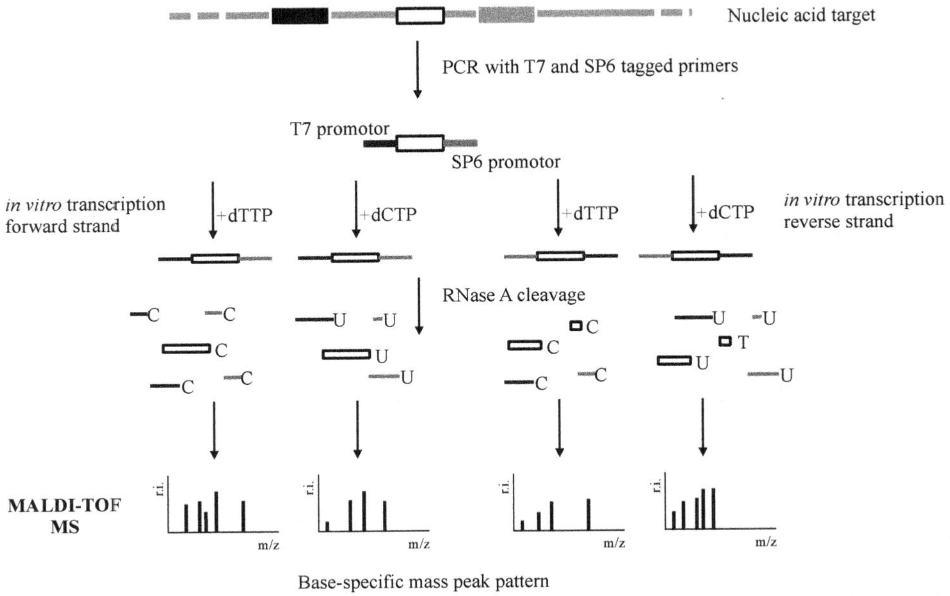

Figure 8.3 Schematic workflow of PCR/base-specific cleavage for comparative sequencing and SNP discovery.

conditioned, desalted and resolved by MALDI-TOF MS. MS spectra for a riboC-specific cleavage of the forward strand, for a riboU-specific cleavage of the forward strand, for a riboC-specific cleavage of the reverse strand and for a riboU-specific cleavage of the reverse strand are acquired. Single-stranded RNA cleavage products (4–30 nucleotides in length) are simultaneously detected in the range of 1000 to 9000 m/z. A time-efficient algorithm matches the resulting mass pattern to *in silico* generated cleavage pattern of a reference-sequence set in a user-generated database. The results are returned with a scoring system and probability call (Honisch *et al.*, 2010). Target regions are identified and sequence polymorphisms can be discovered (Böcker, 2003). A schematic representation of the method is shown in Fig. 8.3. Although this is not a *de novo* sequencing approach, it has numerous applications for resequencing, genotyping and mutation discovery (Ganova-Raeva and Khudyakov, 2013).

The capability of the method to detect multiple nucleic acid target regions in the same run combined with a discriminatory power down to a single nucleotide was first demonstrated for multilocus sequence typing (MLST) of *Neisseria meningitidis*. The concordance with Sanger dideoxy sequencing based MLST was 98.9% (Honisch *et al.*, 2007). MLST uses multiple housekeeping genes to distinguish and relate bacteria on an intra- and interspecies level (Enright and Spratt, 1999). It is a powerful procedure for epidemiological investigations, and has become a gold standard for genotyping of medically relevant bacteria.

Comparative sequencing by MALDI-TOF MS was subsequently applied for the selection of informative gene targets for typing and subtyping of *Salmonella*. The identified set of marker regions has proven to be more discriminatory than the traditional gold standard of serotyping (Bishop *et al.*, 2012).

The identification of a reference sequence combined with the *de novo* detection of single base pair deviations by MALDI-TOF MS was used to identify 13 mutations in *Escherichia coli* K-12 substrain MG1655 during adaptive evolution in the laboratory (Honisch *et al.*, 2004). For the first time subsequent quantification of the identified mutations by MALDI-TOF MS provided monitoring of their fixation during adaption on a whole genome basis (Herring *et al.*, 2006).

Comparative sequencing by MALDI-TOF MS of the *S* gene of Hepatitis B allowed for the automated and accurate identification of all eight virus specific genotypes. Mass peak pattern were automatically compared to sequences from known and database deposited HBV genotypes. Parameters like viral titre, genotype, heterogeneity, quality of PCR and MS pattern were carefully evaluated. The quality of the PCR product was found to have significant impact on the accuracy of the assay. Assay performance again showed concordance with the gold standard dideoxy sequencing results (Ganova-Raeva *et al.*, 2010). Due to its high throughput and semi-automation, the assay was commended for molecular surveillance of HBV infections as a low-cost alternative to dideoxy sequencing (Ganova-Raeva *et al.*, 2012).

Hepatitis C (HCV) is a RNA virus with extremely variable genomic RNA. Its genetic heterogeneity is a hallmark of virus. HCV is classified into six major genotypes and > 50 subtypes (Tellinghuisen *et al.*, 2007). HCV genotypes vary in disease outcome and response to therapy. Several distinct types differ by as much as 33% over the entire genome (Nolte *et al.*, 2003). In each infected individual, HCV exists as multiple variants or quasispecies (Argentini *et al.*, 2009). Consensus sequencing of a hypervariable region (HVR1) and regions in the NS5a and b gene are commonly used to identify HCV transmission. However, a consensus sequence cannot represent the entire HCV population present in the host, particularly in chronically infected patients.

The MALDI-TOF MS sequence specific mass peak pattern of all four concatenated base-specific cleavage reactions were found to reflect the sequence context, heterogeneity and diversity of HCV sample populations and can be mathematically presented as a numeric vector of the detected masses and their corresponding normalized peak intensities. Distances between these vectors have been shown to reflect the genetic relatedness of HCV strains among infected patients for accurate molecular detection of HCV transmission. The mass peak pattern are applicable to evaluating phylogenetic relationships between HCV populations. The approach has been found to match the accuracy of sequence-based approaches, which require time-consuming limited dilution or subcloning to obtain a snapshot of the population structure.

MS pattern recognition and comparison are very promising techniques that can be applied to other pathogens and genomic regions with discriminatory potential to compare linkages between samples or disease states. If no reference sequences are available, RNA cleavage pattern of reference material can be acquired and stored in reference sequence libraries, just as applied for protein based fingerprinting.

The addition of a sodium bisulfite treatment to the target DNA of interest prior to PCR and base-specific cleavage allows for the detection of methylated cytosines in CpG positions of genomic DNA (Ehrlich *et al.*, 2005). Methylation is a central epigenetic process with particular importance for gene regulation, strong implications in the development of cancer and the host response to infectious agents (Deng *et al.*, 2010).

The replacement of gel electrophoresis by MALDI-TOF MS for the separation of

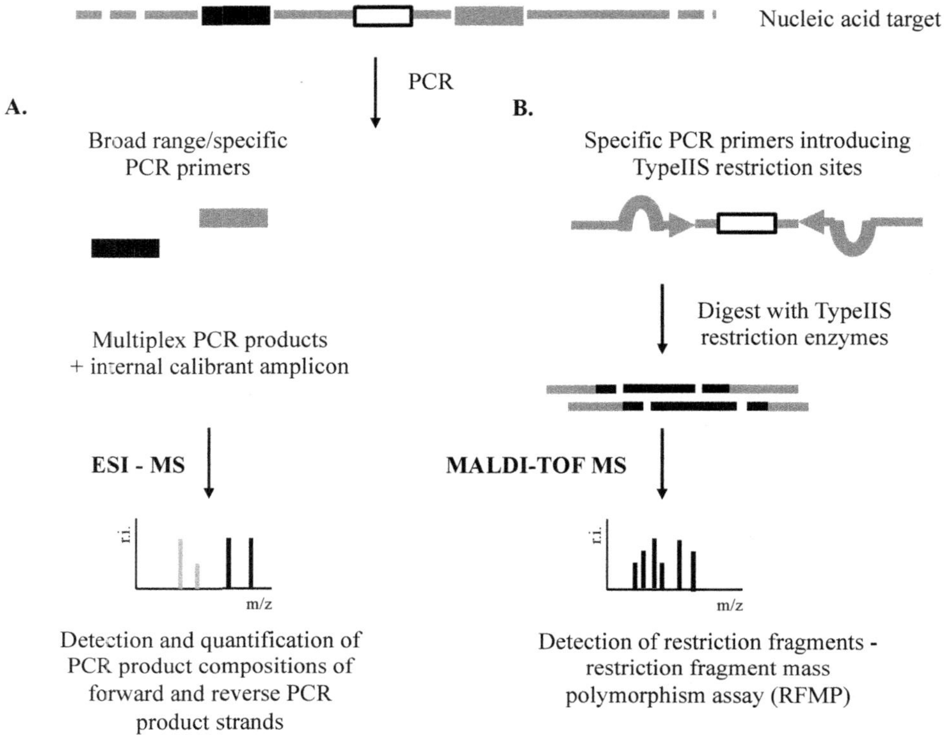

Figure 8.4 Schematic workflow of (A) PCR/ESI MS and (B) RFMP/MALDI-TOF MS.

polymerase chain reaction amplified and restriction digested fragments (PCR-RFLP) was demonstrated for avian influenza viruses (Deyde *et al.*, 2011, Harder *et al.*, 2009), Hepatitis C virus (HCV) (Ilina *et al.*, 2005), drug resistant HBV variants (Han *et al.*, 2011) and Human papillomavirus (HPV) genotyping (Hong *et al.*, 2008). TypeIIS restriction endonuclease recognition sites (e.g. *Fok*I and *Bts*CI) are introduced by PCR surrounding a genotype-specific motif. TypeIIS restriction enzymes cleave DNA at a fixed distance from their recognition site and make the assay independent from restriction sites within the target region of interest. Enzymatic digestion releases a pair of double stranded fragments representative of the genotypic information (Fig. 8.4b). Both strands are analysed by MALDI-TOF MS in parallel providing a level of internal confirmation.

PCR/ESI-MS

The Iridica platform (Abbott), a PCR/ESI-MS technology, uses a targeted phylogenetic approach to microbial, fungal, protozoan and viral identification. Sets of broad range primers in highly conserved regions across bacterial, fungal or protozoan domains of life cover among others 16S rDNA, 23 rDNA, housekeeping gene and resistance gene regions and amplify PCR products of around 100 bp in length, which are after purification/desalting analysed by ESI MS (Fig. 8.4a). The specificity of the method is based on aggregate information from multiple regions of the organisms' genome targeted by a multiplex PCR step. Reference to an internal calibration standard allows in addition to a qualitative answer, if and which target regions are detected, for the quantification of the corresponding analyte. Target identification is accomplished by base composition matching of both PCR strands to a database (Ecker *et al.*, 2008). The principle has found applications in surveillance of infectious diseases (including *Haemophilus influenzae, Neisseria meningitidis, Streptococcus pyrogenes* and *Acinetobacter* spp.) and the detection of known (e.g. *Streptococcus aureus* and *Bacillus anthracis*) and emerging pathogens. Examples are manifold (Ecker *et al.*, 2005, 2006; van Ert *et al.*, 2004).

One study that demonstrates the strength of using a nucleotide composition of an amplicon rather than probing for a specific sequence is the initial identification of the 2009 pandemic H1N1 strain of influenza virus using PCR/ESI-MS. This strain was untypeable and missed using routine targeted molecular methods (Metzgar *et al.*, 2010).

A very interesting multiplex PCR/ESI-MS test that enables a single specimen to be interrogated for the presence of multiple pathogens is the diagnosis of bloodstream infections via analysis of the bacterial domain of life in combination with the detection of genes that mediate antibiotic resistance to certain drugs and virulence factors (Ecker *et al.*, 2010). Methods to rapidly detect and identify pathogens responsible for bloodstream infections, such as septic shock, are desperately needed. Survival rates drop on an hourly basis if appropriate treatments are delayed. Aerobic, anaerobic, culturable, fastidious and unculturable organisms are identified the same way. The platform fulfils some of the relevant features of the ideal molecular diagnostics technology for blood stream infections. Nevertheless, to fully replace the current gold standard of blood culture with a molecular approach like this, clinical studies in comparison to suitable reference methods are necessary and additional resistance mechanisms, which translate into molecular resistance markers, need to be identified. Additional challenges are the specimen preparation and enrichment of pathogen DNA and/or RNA from the background of human genomic DNA in blood or plasma.

The PCR/ESI-MS Iridica platform (Abbott) is currently able to detect and differentiate about 600 bacterial families or fungi and 13 viral families.

Owing to the sample injection step, ESI/MS analysis takes longer than a MALDI-TOF MS analysis as described above. However, the soft ionization process of ESI is particularly suited for the direct detection of large biomolecules such as short PCR products, which makes the technology amenable to a closed-tube PCR only based assay format avoiding a post-PCR step.

Outlook

In general, MS output data can be stored in databases, which allows for access and retrieval of data and their comparison between laboratories, which is important for infectious disease management.

However, although the application of MS to nucleic acid based pathogen detection is used in public health laboratories and vastly described in literature, the mere use of MS as a detection system in place of, for example, real-time PCR or gel electrophoresis does not resonate with clinical laboratories, mainly because of the seemingly high complexity of MS and the cost of the instrumentation. Semi-automation and computational analysis tools to extract clinical information from the complex spectra have, as mentioned before, led to at least one FDA cleared 'moderate-complexity' assay – factor V Leiden and factor II on the Agena IMPACT Dx mass spectrometer.

Nucleic acid MS as well as next-generation sequencing (NGS) have the potential to analyse specimens in a massively parallel fashion and measure molecular details that will enable precise diagnosis of infectious agents. The most immediate use of NGS in clinical microbiology is likely targeted amplification and sequencing as well e.g. for the identification of antibiotic resistance mutations as shown in multidrug-resistant *M. tuberculosis* (Bravo *et al.*, 2009). Yet another advantage of nucleic acid based MS is the multiplexing level of up to 40–60 targets without a fluorescent dye-based channel limitation and the possibility to quantify the analyte. Multiplexing can be beneficial when testing specimens from patients presenting with nonspecific symptoms attributable to a number of different pathogens. Examples include respiratory specimens from patients with suspected viral illness, stool specimen from patients with enteritis, and positive blood cultures as well as infectious disease monitoring in immuno-compromised patients.

MS has the potential to further surpass real-time PCR and electrophoresis in its combined capability of sensitively detecting and sizing of the molecule of interest with an output on the nucleic acid composition in combination with additional analytes like peptides, e.g. from cultured microbial and fungal colonies, lipids and small molecules. If the separation between culture and molecular methods in today's clinical laboratories can be transformed into one centralized laboratory, where a centralized MS platform for multiple analyte detection can be placed, remains to be seen.

References

Alexander, K.A., Laver, P.N., Michel, A.L., Williams, M. van Hilden, P.D., Warren, R.M., and Gey van Pittius, N.C. (2010). Novel *Mycobacterium tuberculosis* complex pathogen, *M. mungi*. Emerg. Infect. Dis. *16*, 1296–1299.

Amexis, G., Oeth, P., Abel, K., Ivshina, A., Pelloquin, F., Cantor, C.R., Braun, A., and Chumakov, K. (2001). Quantitative mutant analysis of viral quasispecies by chip-base matrix-assisted laser desorption/ionization time-of-flight mass spectrometry. Proc. Natl. Acad. Sci. USA *98*, 12097–12102.

Arentini, C., Genovese, D., Dettori, S., and Rapicetta, M., (2009). HCV genetic variability: from quasispecies evolution to genotype classification. Future Microbiol. *4*, 359–373.

Bishop, C., Honisch, C., Goldman, T., Mosko, M., Keng, S., Arnold, C., and Gharbia, S.E. (2012). Combined genomarkers approach to *Salmonella* characterization reveals that nucleotide sequence differences in the phase 1 flagellin gene *fli*C are markers for variation within serotypes. J. Medical Microbiology *61*, 1517–1524.

Böcker, S. (2003). SNP and mutation discovery using base-specific cleavage and MALDI-TOF mass spectrometry. Bioinformatics *19*, 44–53.

Bouakaze, C., Keyser, C., Gonzalez, A., Sougakoff, W., Veziris, N., Dabernat, H., Jaulhac, B., and Ludes, B. (2011). Matrix-assisted laser desorption ionization-time off light mass spectrometry-based single nucleotide polymporphism genotyping assay using iPLEX Gold technology for the identification of *Mycobacterium tuberculosis* complex species and lineages. J. Clin. Microbiol. *49*, 3292–3299.

Bravo, L.T., Tuohy, M.J., Ang, C., Destura, R.V., Mendoza, M., Procop, G.W., Gordon, S.M., Hall, G.S., and Shrestha, N.K. (2009). Pyrosequencing for rapid detection of *Mycobacterium tuberculosis* resistance to rifampin, isoniazid, and fluoroquinolones. J. Clin. Microbiol. *47*, 3985–3990.

Buchan, B.W., and Ledeboer, N.A. (2014). Emerging technologies for the clinical microbiology laboratory. Clin. Microbiol. Rev. *27*, 783–822.

Cobo, F., (2013). Application of MALDI-TOF Mass Spectrometry in clinical virology: a review. Open Virol. J. *7*, 84–90.

Deng, Y.B., Nagae, G., Midorikawa, Y., Yagi, K., Tsutsumi, S., Yamamoto, S., Hasegawa, K., Kokudo, N., Aburatani, H., and Kaneda, A. (2010). Identification of genes preferentially methylated in hepatitis C virus-related hepatocellular carcinoma. Cancer Sci. *10*, 1501–1510.

Deyde, V.M., Sampath, R., and Gubareva, L.V. (2011). RT-PCR/electrospray ionization mass spectrometry approach in detection and characterization of influenza viruses. Expert Rev. Mol. Diagn. *11*, 41–52.

Ecker, D.J., Sampath, R., Blyn, L.B., Eshoo, M.W., Ivy, C., Ecker, J.A., Libby, B., Samant, V., Sannes-Lowery, K.A., Melton, R.E., *et al.* (2005). Rapid identification and strain-typing of respiratory pathogens for epidemic surveillance. Proc. Natl. Acad. Sci. USA *102*, 8012–8017.

Ecker, D.J., Sampath, R., Li, H., Massire, C., Matthews, H.E., Toleno, D., Hall, T.A., Blyn, L.B., Eshoo, M.W., Ranken, R., *et al.* (2010). New technology for rapid molecular diagnosis of bloodstream infections. Expert Rev. Mol. Diagn. *10*, 399–415.

Ecker, D.J., Sampath, R., Massire, C., Blyn, L.B., Hall, T.A., Eshoo, M.W., and Hofstadler, S.A. (2008). Ibis T5000: a universal biosensor approach for microbiology. Nat. Rev. Microbiol. *6*, 553–558.

Ecker, J.A., Massire, C., Hall, T.A., Ranken, R., Pennella, T.D., Agasino, C., Blyn, L.B., Hofstadler, S.A., Endy, T.P., Scott, P.T., *et al.* (2006). Identification of Acinetobacter species and genotyping of *Acinetobacter baumannii* by multilocus PCR and mass spectrometry. J. Clin. Microbiol., *44*, 2921–2932.

Ehrlich, M., Nelson, M.R., Stanssens, P., Zabeau, M., Liloglou, T., Xinarianos, G., Cantor, C.R., Field, J.K., van den Boom, D. (2005). Quantitative high-throughput analysis of DNA methylation pattern by base-specific cleavage and mass spectrometry. Proc. Nat. Acad. Sci. USA *102*, 15785–15790.

Enright, M.C., and Spratt, B.G. (1999). Multilocus sequence typing. Trends Microbiol. 7, 482–487.

Fenn, J.B., Mann, M., Meng, C.K., Wong, S.F., and Whitehouse, C.M. (1989). Electrospray ionization for mass spectrometry of large biomolecules. Science *246*, 64–71.

Ganova-Raeva, L.M., Dimitrova, Z.E., Campo, D.S., and Khudyakov, Y.E. (2012). Application of mass spectrometry to molecular surveillance of hepatitis B and C viral infections. Antivir. Ther. *17*, 1477–1482.

Ganova-Raeva, L.M., Dimitrova, Z.E., Campo, D.S., Lin, Y., Ramachandran, S., Xia, G., Honisch, C., Cantor, C.R., and Khudyakov, Y.E. (2013). Detection of hepatitis C virus transmission by use of DNA mass spectrometry. J. Infect. Dis. *207*, 999–1006.

Ganova-Raeva, L.M., and Khudyakov, Y.E. (2013). Applications of mass spectrometry to molecular diagnostics of viral infections. Expert Rev. Mol. Diagn. *13*, 377–388.

Ganova-Raeva, L.M., Ramachandran, S., Honisch, C., Forbi, J.C., Zhai, X., and Khudyakov, Y. (2010). Robust hepatitis B virus genotyping by mass spectrometry. J. Clin. Microbiol. *48*, 4161–4168.

Han, K.H., Hong, S.P., Choi, S.H., Shin, S.K., Cho, S.W., Ahn, S.H., Hahn, J.S., and Kim, S.O. (2011). Comparison of multiplex restriction fragment mass polymorphism and sequencing analysis for detecting entecavir resistance in chronic hepatitis B. Antivir. Ther. *16*, 77–87.

Harder, M.K., Mettenleiter, T.C., and Karger. A. (2009). Diagnosis and strain differentiation of avian influenza viruses by restriction fragment mass analysis. J. Virol. Methods *158*, 63–69.

Herring, C.D., Honisch, C., Raghunathan, A., Patel, T., Applebee, M.K., Joyce, A.R., Albert, T.J., Blattner, F.R., van den Boom, D., Cantor, C.R., *et al.* (2006). Comparative genome sequencing of *Escherichia coli* allows observation of bacterial evolution on a laboratory timescale. Nat. Gen. *38*, 1406–1412.

Hong, S.P., Shin, S.K., Lee, E.H., Kim, E.O., Ji, S.I., Chung, H.J., Park, S.N., Yoo, W., Folk, W.R., and Kim, S.O. (2008). High-resolution human papillomavirus genotyping by MALDI-TOF mass spectrometry. Nature Protocols 3, 1476–1484.

Honisch, C., Chen, Y., and Hillenkamp, F. (2010). Comparative DNA sequence analysis and typing using mass spectrometry. In Mass Spectrometry for Microbial Proteomics, Shah, H.N., and Gharbia, S.E., eds. (John Wiley and Sons Ltd), pp. 441–462.

Honisch, C., Chen, Y., Mortimer, C., Arnold, C., Schmidt, O., van den Boom, D., Cantor, C.R., Sha, H.N., and Gharbia, S.E. (2007). Automated comparative sequence analysis by base-specific cleavage and mass spectrometry for nucleic-acid-based microbial typing. Proc. Natl. Acad. Sci. USA *104*, 10649–10654.

Honisch, C., Mosko, M., Arnold, C., Gharbia, S.E., Diel, R., and Niemann, S. (2010). Replacing reverse line blot hybridization spoligotyping of the *Mycobacterium tuberculosis* complex. J. Clin. Microbiol. *48*, 1520–1526.

Honisch, C., Raghunathan, A., Cantor, C.R., Palsson, B.Ø., and van den Boom, D. (2004). High-throughput mutation detection underlying adaptive evolution of *Escherichia coli*-K12. Genome Res. *14*, 2495–2502.

Ikryannikova, L.N., Afanas'ev, M.V., Akopain, T.A., Ill'ina, E.N., Kuz'min, A.V., Larionova, E.E., Smirnova, T.G., Chernousova, L.N., and Govorun, V.M. (2007). Mass-spectrometry based minisequencing method for the rapid detection of drug resistance in *Mycobacterium tuberculosis*. J. Clin. Microbiol. Methods *70*, 395–405.

Ilina, E.N., Malakhova, M.V., Generozov, E.V., Nikolaev, E.N., and Govorun, V.M. (2005). Matrix-assisted laser desorption ionization-time of flight (mass spectrometry) for hepatitis C virus genotyping. J. Clin. Microbiol. *43*, 2810–2815.

Jurinke, C., Oeth, P., and van den Boom, D. (2004). MALDI-TOF mass spectrometry: a versatile tool for high-performance DNA analysis. Mol. Biotechnol. *26*, 147–164.

Kamerbeek, J., Schouls, L., Kolk, A., van Agterveld, M., van Soolingen, D., Kuijper, S., Bunschoten, A., Molhuizen, H., Shaw, R., Goyal, M., *et al.* (1997). Simultaneous detection and strain differentiation of *Mycobacterium tuberculosis* for diagnosis and epidemiology. J. Clin. Microbiol. *35*, 907–914.

Karas, M., and Hillenkamp, F. (1988). Laser desorption ionization of proteins with molecular masses exceeding 10,000 daltons. Anal. Chem. *60*, 2299–2301.

Kirpekar, F., Nordhoff, E., Kristiansen, K., Roepstorff, P., Lezius, A., Hahner, S., Karas, M., and Hillenkamp, F. (1994). Matrix assisted laser desroption/ionization mass spectrometry of enzymatically synthesized RNA up to 150 kDa. Nucleic Acids Res. *22*, 3866–3870.

Köster, H., Tang, K., Fu, D.J., Braun, A., van den Boom, D., Smith, C.L., Cotter, R.J., and Cantor, C.R. (1996). A strategy for rapid and efficient DNA sequencing by mass spectrometry. Nat. Biotechnol., *14*, 1123–1128.

Liu, J., Lim, S.L., Ruan, Y., Ling, A.E., Ng, L.F.P., Drosten, C., Liu, E.T., Stanton, L.W., and Hibberd, M.L. (2005). SARS transmission pattern in Singapore reassessed by viral sequence variation analysis. PLoS Medicine 2, e43.

Maxam, A.M., and Gilbert, W. (1977). A new method for sequencing DNA. Proc. Natl. Acad. Sci. USA *74*, 560–564.

Metzgar, D., Baynes, D., Myers, C.A., Kammerer, P., Unabia, M., Faix, D.J., and Blair, P.J. (2010). Initial identification and characterization of an emerging zoonotic influenza virus prior to pandemic spread. J. Clin. Microbiol. *48*, 4228–4234.

Nolte, F.S.A., Green, M.K., Fiebelkorn, R., Caliendo, A.M., Sturchio, C., Grunwald, A., and Healy, M. (2003). Clinical evaluation of two methods for genotyping hepatitis C virus based on analysis of the 5′ noncoding region. J. Clin. Microbiol. *41*, 1558–1564.

Nordhoff, E., Cramer, R., Karas, M., Hillenkamp, F., Kirpekar, F., Kristiansen, K., and Roepstorff, P. (1993). Ion stability of nucleic acids in infrared matrix-assisted laser desorption/ionization mass spectrometry. Nucleic Acid Res. *21*, 3347–3357.

Nordhoff, E., Ingendoh, A., Cramer, R., Overberg, A., Stahl, B., Karas, M., Hillenkamp, F., and Rothschild, K.J. (1992). Matrix-assisted laser desorption/ionization mass spectrometry of nucleic acids with wavelengths in the ultraviolet and infrared. Rapid Commun. Mass Spectrom. *6*, 771–776.

Nygren, A., Mosko, M., Flores, E., Nakorchevsky, A., Ehrich, M., Honisch, C., and van den Boom, D. (2013). A novel approach for multiplex ultrasensitive detection of somatic mutations in tumor samples. Poster presented at American Association for Cancer Research (Philadelphia, PA).

Piao, J., Jiang, J., Xu, B., Wang, X., Guan, Y., Wu, W., Liu, L., Zhang, Y., Huang, X., Wang, P., *et al.* (2012). Simultaneous detection and identification of eneric viruses by PCR Mass-Array. PLoS One 7, 42251e.

Ramaswamy, S.V., and Musser, J.M. (1998). Molecular genetic basis of antimicrobial agent resistance in *Mycobacterium tuberculosis*. Tuber. Lung Dis. *79*, 3–29.

Ramaswamy, S.V., Reich, R., Dou, S.J., Jasperse, L., Pan, X., Wanger, A., Quitugua, T., and Graviss, E.A. (2003). Single nucleotide polymorphisms in genes associated with isoniazid resistance in *Mycobacterium tuberculosis*. Antimicrob. Agents Chemother. 47, 1241–1250.

Sauer, S., and Kliem, M. (2010). Mass spectrometry tools for the classification and identification of bacteria. Nat. Rev. Microbiol. *8*, 74–82.

Soini, H., and Musser, J.M. (2001). Molecular diagnosis of *mycobacteria*. Clin. Chem. *47*, 809–814.

Stanssens, P., Zabeau, M., Meerseman, G., Remes, B., Gansemans, Y., Storm, N., Hartmer, R., Honisch, C., Rodi, C.P., Böcker, S., *et al.* (2004). High-throughput MALDI-TOF discovery of genomic sequence polymorphisms. Genome Res. *14*, 126–133.

Stenmark, M.H., McHugh, J.B., Schipper, M., Walline, H.M., Komarck, C., Feng, F.Y., Worden, F.P., Wolf, G.T., Chepeha, D.B., Prince, M.E., *et al.* (2013). Nonendemic HPV-positive nasopharyngeal carcinoma: association with poor prognosis. Int. J. Radiation Oncol. Biol. Phys., *88*, 580–588.

Storm, N.B., Darnhofer-Patel, B., van den Boom, D., and Rodi, C.P. (2002). MALDI-TOF mass spectrometry-based SNP genotyping. Methods Mol. Biol. *212*, 241–262.

Sjöholm, M.I.L., Dillner, J., and Carlson, J. (2008). Multiplex detection of human herpesviruses from archival specimens by using matrix-assisted laser desorption ionization-time of flight mass spectrometry. J. Clin. Microbiol. *46*, 540–545.

Syrmis, M.W., Kidd, T.J., Moser, R.J., Ramsay, K.A., Gibson, K.M., Anuj, S., Bell, S.C., Wainwright, C.E., Grimwood, K., Nissen, M., *et al.* (2014). A comparison of two informative SNP-based strategies for typing *Pseudomonas aeruginosa* isolates from patients with cystic fibrosis. BMC Infectious Diseases *14*, 307–315.

Symis, M.W., Moser, R.J., Whiley, D.M., Vaska, V., Coombs, G.W., Nissen, M.D., Sloots, T.P., and Nimmo, G.R. (2011). Comparison of a multiplexed MassARRAY system with real-time allele-specific PCR technology for genotyping of methicillin-resistant *Staphylococcus aureus*. Clin. Microbiol. Infect. *17*, 1804–1810.

Tang, K., Fu, D.J., Kotter, S., Cotter, R.J., Cantor, R.J., and Köster, H. (1995). Matrix-assisted laser desorption/ionization mass spectrometry of immobilized duplex DNA probes. Nucleic Acid Res. *23*, 3126–3131.

Tellinghuisen, T.L., Evans, M.J., von Hahn, T., You, S., and Rice, C.M. (2007). Studying hepatitis C virus: Making the best of a bad virus. J. Virol. *81*, 8853–8867.

Trembizki, E., Smith, H., Monica, M., Lahra, M., Chen, M., Donovan, B., Fairley, C.K., Guy, R., Kaldor, J., Degan, D., *et al.* (2014). High-throughput informative single nucleotide polymporphism-based typing of *Neisseria gonorrhoeae* using the Sequenom MassARRAY iPLEX platform. J. Antimicrob. Chemother. *69*, 1526–1532.

Tsukamura, M., Mizuno, S., and Toyama, H. (1985). Taxonomic studies on the *Mycobacterium tuberculosis* series. *29(4)*, 285–299.

van Belkum, A., Welker, M., Erhard, M., and Chatellier, S. (2012). Biomedical mass spectrometry in today's and tomorrow's clinical microbiology laboratories. J. Clin. Microbiol. *50*, 1513–1517.

Van Ert, M.N., Hofstadtler, S.A., Jiang, Y., Busch, J.D., Wagner, D.M., Drader, J.J., Ecker, D.J., Hannis, J.C., Huynh, L.Y., Schupp, J.M., *et al.* (2004). Mass spectrometry provides accurate characterization of two genetic marker types in *Bacillus anthracis*. BioTechniques *37*, 642–651.

Vivekanandan, P., Daniel, H.D., Kannangai, R., Martinez-Murillo, F., and Torbenson, M. (2010). Hepatitis B virus replication induces methylation of both host and viral DNA. J. Virol. *84*, 4321–4329.

von Witzingerode, F., Boecker, S., Schloetelburg, C., Chiu, N.H., Storm, N., Jurinke, C., Cantor, C.R., Goebel, U.B., and van den Boom, D. (2002). Base-specific fragmentation of amplified 16S rRNA genes analyzed by mass spectrometry: a tool for rapid bacterial identification. Proc. Natl. Acad. Sci. USA *99*, 7039–7044.

Wu, K.J., Steding, A., and Becker, C.H. (1993). Matrix-assisted laser desorption time-of-flight mass spectrometry of oligonucleotides using 3-hydroxy-picolinic acid as an ultraviolet-sensitive matrix. Rapid Commun. Mass Spectrom. 7, 142–146.

Yang, H., Yang, K., Khafagi, A., Tang, Y., Carey, T.E., Opipari, A.W., Lieberman, R., Oeth, P.A., Lancaster, W., Klinger, H.P., *et al.* (2004). Sensitive detection of human papillomavirus in cervical, head/neck, and schistosomiasis-associated bladder malignancies. Proc. Natl. Acad. Sci. USA *102*, 7683–7688.

Future Trends and Perspectives of MALDI-TOF MS in the Microbiology Laboratory

9

Sören Schubert and Markus Kostrzewa

Abstract

Within less than a decade matrix-assisted laser desorption/ionization time-of-flight mass spectrometry (MALDI-TOF MS) has become a gold standard for microbial identification in clinical microbiology laboratories. Besides identification of microorganisms the typing of single strains as well as the antibiotic and antimycotic resistance testing has come into focus in order to speed up the microbiological diagnostic. However, the full potential of MALDI-TOF MS has not been tapped yet and future technological advancements will certainly expedite this method towards novel applications and enhancement of current practice. So, the following chapter shall be rather a brainstorming and forecast of how MALDI-TOF MS will develop to influence clinical diagnostics and microbial research in the future. It shall open up the stage for further discussions and does not claim for overall validity.

Introduction

Within less than a decade MALDI-TOF MS has entered the microbiological diagnostic laboratories around the world providing a fast, cheap and reliable tool for identification of bacteria and fungi cultivated on agar plates or in liquid media. First steps have been made to introduce MALDI-TOF MS for antibiotic and antimycotic resistance testing as well as for some typing applications. The future development of MALDI-TOF MS for clinical diagnostics will likely be driven by technical advances, on the machine and software side as well as regarding sample preparation, and integration of MALDI-TOF MS in fully automated workflows pushing forward already established application. On the other hand, new fields of application for MALDI-TOF MS may be envisaged extending the use on imaging and maybe even *in vivo* applications. This development should bring together different laboratory disciplines like pathologists, medical microbiologist and pharmacologists.

Improvement of current MALDI-TOF MS technologies

One of the foreseeable main developments for microbiological laboratories in the next future will be the implementation of automated processes in the lab workflow, in particular in diagnostic laboratories. Several diagnostic companies, e.g. BD Kiestra (Drachten, The Netherlands) and Copan (Brescia, Italy) have developed fully automated workflows for the culture based microbiological laboratories (Matthews and Deutekom, 2011). Here, the allocation and barcode labelling of agar plates, streaking of the samples onto the plates and inoculation of liquid culture media as well as the transfer of culture media to specialized incubators is managed by a fully automated robotic system. Moreover, growth on the plates can be monitored and documented by digital imaging at defined time points. The complete implementation of MALDI-TOF MS in these workflows, however, is still in its infancy as several steps in the 'from-the sample-to-result' workflow are still done manually. Today, the identification using MALDI-TOF MS is performed on demand initiated by lab personnel at different time points. The selection of respective colonies for further identification is essentially based on the skills and experience of the staff. In future, a fully automated, seamless MALDI-TOF MS workflow for identification could be triggered by software assessing the images of the agar plate for putatively different colonies. These colonies then could be prepared automatically for the MALDI-TOF measurement. Subsequently, sample targets might be transferred to and introduced into the mass spectrometer in an automated manner, without human interaction. Depending on the identification results microbouillons could be inoculated by robotics to initiate antibiotic resistance testing. From this, the system could be adjusted to perform further tests like resistance testing only from previously defined bacterial species in case specimen from mucosal surfaces are investigated. In samples from primarily sterile body-sites all growing microorganisms would be selected. The new generation of MALDI-TOF MS equipped with high-performance lasers (up to 10.000 laser shots per second versus 200 Hz operating in the fastest commercial microbial ID MALDI-TOF MS available today) gives a glimpse of how powerful the future technical development of MALDI-TOF MS will be. As soon as rapid MALDI-TOF MS based resistance testing and integrated automated typing of resistance or virulence markers will enter the stage, a further significant step towards speeding up the diagnostic workflows in microbiological laboratories will be in our grasps.

Immuno-MALDI and imaging mass spectrometry: potential for further value of MALDI-TOF MS in microbiology

The application of MALDI-TOF MS as a standard analysis method in clinical routine diagnostics beyond microbial identification has not yet been established. On the other hand, in view of a deeper understanding of physiological processes and the discovery of new biomarkers, the requirements for sophisticated analysis methods are growing. Additionally, modern disease management demands for better diagnostic tools to improve early diagnostics and monitoring of therapeutic intervention. Established immunological methods for biomarker detection and quantification, i.e. ELISA (enzyme-linked immunosorbent assay) or RIA (radio immune assay) rely on the availability of two antibodies, one to capture the target molecule and the second for signal generation, and use a single principle of detection for both selective steps. Thereby, these conventional detection methods do not reveal any further molecular information. The combination of immunoaffinity and MALDI-TOF MS, Immuno-MALDI, may provide much better specificity because the antibody is only used for antigen capturing whereas the MALDI-TOF analysis determines an intrinsic characteristic of the respective molecule of interest, i.e. its molecular weight. Thus, the specificity of analysis can be significantly improved by using a capture antibody combined with specific mass identification by MALDI-TOF (Sparbier *et al.*, 2009). This can be of particular advantage in case small proteins and peptides are in the focus, which hardly offer an epitope for a second antibody. This holds also true if secondary modified molecules have to be differentiated, e.g.

glycosylated or phosphorylated molecules. Quantification can be achieved by spiking and co-capture of stable-isotope labelled, heavier peptides of exactly the same composition. For these labelled peptides a calibration curve has to be established. Development of methods to quantify hypertension markers has already been described (Reid *et al.*, 2010; Camenzind *et al.*, 2013). Applications of this technology to solve microbiology-related problems, e.g. toxin quantification, still have to be demonstrated.

Another MALDI-TOF MS based technology, imaging mass spectrometry (IMS), has been recently developed and might be applicable in diagnostic laboratories in the near future. This imaging technique allows molecular mapping of different kind of biomolecules in their natural environment (Walch *et al.*, 2008; Neubert and Walch, 2013; Aichler and Walch, 2015). IMS has already entered the field of tissue-based research providing unique advantages for analysing tissue specimen in an unprecedented detail. As a mass spectrometry-based technology combined with the two-dimensional mapping of molecules it enables the direct correlation of tissue histology and proteomic, metabolomic or lipidomic information. IMS allows for a label-free analysis of numerous analytes of multiple types. Most encouragingly for clinical purposes, this technology keeps the tissue intact, thereby allowing investigation of tissue morphology by traditional clinical standard procedures, in parallel (Neubert and Walch, 2013). IMS can analyse a multitude of analytes ranging from proteins, peptides, protein modifications to small molecules, drugs and their metabolites as well as pharmaceutical components, endogenous cell metabolites, lipids, and other analytes in situ, without any labelling. While current research and development in this area is mainly focused on cancer diagnostics and inflammatory diseases like inflammatory bowel disease (M'Koma, 2014; Kriegsmann *et al.*, 2015), IMS might also have a future role in microbiology research and even diagnostic. Direct profiling of molecules from tissue samples by MALDI-TOF MS provides a means to study the pathogen–host interaction and to discover potential markers of infection (Moore *et al.*, 2014a,b). Further, IMS is able to investigate metabolic exchange factors of intraspecies, interspecies, and polymicrobial interactions (Yang *et al.*, 2012). Although still in its infancy, the application of imaging mass spectrometry in microbiology therefore might allow for fundamentally new insights into microbial communities.

The future will show in which way MALDI-TOF MS is going to further revolutionize microbiology diagnostics.

References

Aichler, M., and Walch, A. (2015). MALDI Imaging mass spectrometry: current frontiers and perspectives in pathology research and practice. Lab. Invest. *95*, 422–431.

Balluff, B., Rauser, S., Ebert, M.P., Siveke, J.T., Hofler, H., and Walch, A. (2012). Direct molecular tissue analysis by MALDI imaging mass spectrometry in the field of gastrointestinal disease. Gastroenterology *143*, 544–549.

Burnum, K.E., Frappier, S.L., and Caprioli, R.M. (2008). Matrix-assisted laser desorption/ionization imaging mass spectrometry for the investigation of proteins and peptides. Annu. Rev. Anal. Chem. (Palo. Alto. Calif.) *1*, 689–705.

Camenzind, A.G., van der Gugten, J.G., Popp, R., Holmes, D.T., and Borchers, C.H. (2013). Development and evaluation of an immuno-MALDI (iMALDI) assay for angiotensin I and the diagnosis of secondary hypertension. Clin. Proteomics *10*, 20.

Fournier, I., Wisztorski, M., and Salzet, M. (2008). Tissue imaging using MALDI-MS: a new frontier of histopathology proteomics. Expert. Rev. Proteomics *5*, 413–424.

Germain, R.N. (2005). Imaging dynamic interactions in immune responses. Semin. Immunol. *17*, 385–386.

Gessel, M.M., Norris, J.L., and Caprioli, R.M. (2014). MALDI imaging mass spectrometry: spatial molecular analysis to enable a new age of discovery. J. Proteomics *107*, 71–82.

Kriegsmann, J., Kriegsmann, M., and Casadonte, R. (2015). MALDI TOF imaging mass spectrometry in clinical pathology: a valuable tool for cancer diagnostics (review). Int. J. Oncol. *46*, 893–906.

Matthews, S., and Deutekom, J. (2011). The future of diagnostic bacteriology. Clin. Microbiol. Infect. *17*, 651–654.

M'Koma, A.E. (2014). Diagnosis of inflammatory bowel disease: potential role of molecular biometrics. World J. Gastrointest. Surg. *6*, 208–219.

Moore, J.L., Becker, K.W., Nicklay, J.J., Boyd, K.L., Skaar, E.P., and Caprioli, R.M. (2014a). Imaging mass spectrometry for assessing temporal proteomics: analysis of calprotectin in *Acinetobacter baumannii* pulmonary infection. Proteomics *14*, 820–828.

Moore, J.L., Caprioli, R.M., and Skaar, E.P. (2014b). Advanced mass spectrometry technologies for the study of microbial pathogenesis. Curr. Opin. Microbiol. *19*, 45–51.

Neubert, P., and Walch, A. (2013). Current frontiers in clinical research application of MALDI imaging mass spectrometry. Expert. Rev. Proteomics *10*, 259–273.

Reid, J.D., Holmes, D.T., Mason, D.R., Shah, B., and Borchers, C.H. (2010). Towards the development of an immuno MALDI (iMALDI) mass spectrometry assay for the diagnosis of hypertension. J. Am. Soc. Mass Spectrom. *21*, 1680–1686.

Sparbier, K., Wenzel, T., Dihazi, H., Blaschke, S., Muller, G.A., Deelder, A., Flad, T., and Kostrzewa, M. (2009). Immuno-MALDI-TOF MS: new perspectives for clinical applications of mass spectrometry. Proteomics *9*, 1442–1450.

Walch, A., Rauser, S., Deininger, S.O., and Hofler, H. (2008). MALDI imaging mass spectrometry for direct tissue analysis: a new frontier for molecular histology. Histochem. Cell Biol. *130*, 421–434.

Yang, J.Y., Phelan, V.V., Simkovsky, R., Watrous, J.D., Trial, R.M., Fleming, T.C., Wenter, R., Moore, B.S., Golden, S.S., Pogliano, K., *et al.* (2012). Primer on agar-based microbial imaging mass spectrometry. J. Bacteriol. *194*, 6023–6028.

Index

D

E

N

Lightning Source UK Ltd.
Milton Keynes UK
UKOW07f0344201216
290456UK00003B/9/P

9 781910 190418